高职高专院校"十三五"精品示范系列教材（软件技术专业群）

Java ME 手机应用程序开发

易　灿　李志勇　编　著

刘彦姝　主　审

中国水利水电出版社
www.waterpub.com.cn
·北京·

内 容 提 要

本书通过一些实际案例详细地讲述了 Java ME 手机游戏开发的开发环境、基础知识和相关开发技术以及手机游戏编程的基本思想和思路。

本书共十章,分三个模块设计。第一个模块为入门的基础知识讲解,包括第 1、2 章;第二个模块为 Java ME 理论基础知识的应用,包括图形用户界面、相关组件、信息保存等综合内容,并辅以一个小实训项目来具体分析,本模块涵盖第 3、4、5、6 章;第三个模块为 Java ME 高级应用模块,在这个模块里面,通过一个市面上常见的手机游戏类型的综合项目阐述,让读者能基本掌握 Java ME 手机游戏开发的相关技术和开发流程,本模块主要包括第 7、8、9、10 章。

本书适合高职高专软件技术、计算机应用及相关专业学生学习使用,也适合作为具有其他语言或者平台游戏开发经验并且想使用 Java ME 开发游戏的游戏开发者参考。

图书在版编目(CIP)数据

Java ME 手机应用程序开发 / 易灿,李志勇编著
. -- 北京:中国水利水电出版社,2017.3
高职高专院校"十三五"精品示范系列教材. 软件技术专业群
ISBN 978-7-5170-5102-2

Ⅰ. ①J… Ⅱ. ①易… ②李… Ⅲ. ①JAVA语言-程序设计-高等职业教育-教材②移动通信-携带电话机-应用程序-程序设计-高等职业教育-教材 Ⅳ.
①TP312.8②TN929.53

中国版本图书馆CIP数据核字(2017)第013560号

责任编辑:周益丹   加工编辑:陈宏华   封面设计:李 佳

书 名	高职高专院校"十三五"精品示范系列教材(软件技术专业群) Java ME 手机应用程序开发 Java ME SHOUJI YINGYONG CHENGXU KAIFA
作 者	易 灿 李志勇 编 著 刘彦姝 主 审
出版发行	中国水利水电出版社 (北京市海淀区玉渊潭南路 1 号 D 座 100038) 网址:www.waterpub.com.cn E-mail:mchannel@263.net(万水) 　　　　sales@waterpub.com.cn 电话:(010)68367658(营销中心)、82562819(万水)
经 售	全国各地新华书店和相关出版物销售网点
排 版	北京万水电子信息有限公司
印 刷	北京瑞斯通印务发展有限公司
规 格	184mm×240mm 16 开本 16.25 印张 357 千字
版 次	2017 年 3 月第 1 版 2017 年 3 月第 1 次印刷
印 数	0001—3000 册
定 价	34.00 元

凡购买我社图书,如有缺页、倒页、脱页的,本社营销中心负责调换

版权所有·侵权必究

前 言

每一个人都会玩游戏，都喜欢玩游戏；但并不是每一个人都会开发制作游戏。要让游戏无处不在，还需要更多的人学会开发游戏。相信好学的你在玩过几个好玩的游戏后也会问：这个游戏是怎么做出来的？我是否也可以做出同样出色的游戏？我该如何学习游戏的制作？……游戏有很多种——本书中会讲到——在当今的2016年如果想找一种游戏，它只需要一个人业余花费很少的时间和精力就可以制作出，而且有可能会成为非常受欢迎的游戏，我想非移动游戏莫属！

Java ME是Sun公司提供的移动应用开发平台。自从Sun公司发布Java ME以来，Java ME技术便引起了软件开发商、信息服务商的极大关注，超过500家公司签订了使用Java ME的协议。主要的移动设备制造商，如诺基亚、西门子、三星、摩托罗拉等公司都推出了支持Java ME技术的手机。现在，有越来越多的人意识到了Java ME技术的开发与应用带来的无限机遇。本书主要面向有一定Java基础的开发人员和高校学生。

本书作为Java ME移动游戏（主要是手机游戏）制作的入门读物，只要你具备Java编程的基础知识并且了解一些Java ME的背景知识，通过本书的学习，你就能开发出自己的游戏。当然，最最重要的还是读者你的创意！

本书由三个模块组成，下面分别介绍：

第一个模块为基础模块，包含第1~2章，本模块主要介绍Java ME手机游戏开发相关的基础知识和编程环境搭建。本模块主要由易灿编写。

第二个模块为基础应用模块，本模块主要介绍Java ME在游戏开发方面应用的相关知识。其中第3、4章由李志勇编写，第5、6章由刘彦姝编写。

第三个模块包含第7、8、9、10章，本模块主要通过开发一个完整的手机游戏来阐述Java ME手机游戏开发的整体思想和相关技术的应用，由易灿编写。

本书的全部代码均在JDK1.6+WTK2.2环境下调试通过，并在WTK自带的模拟器上能够正确运行。本书代码仅供学习Java ME手机游戏开发的编程人员和学习者使用，欢迎读者对书中不当之处提出批评建议。

本书是国家骨干高等职业院校重点建设项目研究成果之一，由易灿、李志勇编著，刘彦姝主审，适用读者对象是高职高专软件技术、计算机应用及相关专业学生，也可作为具有其他语言或者平台游戏开发经验并且想使用 Java ME 开发游戏的游戏开发者参考。

编　者
2016 年 12 月

目 录

前言

第 1 章　Java ME 概述 1
- 1.1　Java ME 体系结构 1
  - 1.1.1　Java 的版本 1
  - 1.1.2　Java ME 的 3 层体系结构 2
  - 1.1.3　虚拟机（KVM） 3
- 1.2　有限连接设备配置表（CLDC） 4
  - 1.2.1　CLDC 概览 4
  - 1.2.2　CLDC 中使用的 J2SE 类 5
  - 1.2.3　CLDC 专用类 8
  - 1.2.4　CLDC 1.1 的新特性 9
- 1.3　MIDP 11
  - 1.3.1　设备需求 11
  - 1.3.2　MIDP 的总体体系结构 13
  - 1.3.3　MIDP 类库 14
  - 1.3.4　MIDP 2.0 的新特性 15
  - 1.3.5　MIDP 2.0 的安全机制 16
- 1.4　本章小结 18

第 2 章　搭建开发平台——Eclipse 19
- 2.1　初识 Eclipse、EclipseME、WTK 19
  - 2.1.1　Eclipse 19
  - 2.1.2　EclipseME 20
  - 2.1.3　其他工具和环境 20
- 2.2　搭建 Eclipse 移动开发环境 20
  - 2.2.1　安装 JDK 1.6 20
  - 2.2.2　安装 Eclipse 22
  - 2.2.3　安装 EclipseME 插件 23
- 2.3　加载厂商模拟器 24
- 2.4　Java ME 项目开发 25
  - 2.4.1　创建工程 25
  - 2.4.2　创建 Midlet 类 27
  - 2.4.3　执行 Midlet 29
  - 2.4.4　打包与混淆 30
- 2.5　本章小结 30

第 3 章　MIDP 高级 UI 的使用 31
- 3.1　概述 31
- 3.2　列表 List 32
  - 3.2.1　Exclusive（单选式） 33
  - 3.2.2　Implicit（隐含式） 33
  - 3.2.3　Multiple（多选式） 34
- 3.3　TextBox 36
- 3.4　Alert 39
- 3.5　Form 概述 43
- 3.6　StringItem 及 ImageItem 44
  - 3.6.1　StringItem 44
  - 3.6.2　ImageItem 46
- 3.7　CustomItem 47
- 3.8　TextField 和 DateField 55
- 3.9　Gauge 和 Spacer，ChoiceGroup 56

3.9.1　Gauge……………………… 56	5.5　本章小结……………………………… 86
3.9.2　Spacer……………………… 58	第6章　GAME API（MIDP2.0）………… 87
3.9.3　ChoiceGroup……………… 58	6.1　游戏API简介………………………… 87
3.10　本章小结……………………………… 58	6.2　GameCanvas的使用………………… 88
第4章　MIDP低级UI的使用……………… 59	6.2.1　绘图……………………… 89
4.1　低级API与低级事件响应…………… 60	6.2.2　键盘……………………… 90
4.2　重绘事件及Graphics………………… 61	6.3　Sprite的使用………………………… 90
4.2.1　坐标概念…………………… 61	6.3.1　Sprite帧………………… 91
4.2.2　颜色操作…………………… 61	6.3.2　帧序列…………………… 91
4.2.3　绘图操作…………………… 62	6.3.3　ReferencePixel…………… 93
4.3　Canvas与屏幕事件处理……………… 65	6.3.4　Sprite的变换……………… 94
4.4　键盘及触控屏幕事件的处理………… 67	6.3.5　绘制Sprite………………… 95
4.5　Graphics相关类……………………… 69	6.3.6　碰撞检测………………… 95
4.5.1　Image类…………………… 69	6.4　Layer的使用………………………… 96
4.5.2　字体类……………………… 73	6.4.1　TiledLayer………………… 96
4.6　本章小结……………………………… 74	6.4.2　LayerManager…………… 98
第5章　MIDP的数据存储——RMS……… 75	6.5　一个示例……………………………… 100
5.1　初识RMS（Record Management System）……………………………… 75	6.6　本章小结……………………………… 116
	第7章　手机RPG游戏设计与实现……… 117
5.2　RecordStore的管理………………… 76	7.1　游戏概述……………………………… 117
5.2.1　RecordStore的打开……… 76	7.2　游戏启动画面………………………… 118
5.2.2　RecordStore的关闭……… 77	7.3　游戏主菜单的实现…………………… 120
5.2.3　RecordStore的删除……… 78	7.4　"关于我们"菜单的实现…………… 124
5.2.4　其他相关操作……………… 78	7.5　"游戏帮助"菜单的实现…………… 126
5.3　RecordStore的基本操作…………… 79	7.6　"游戏设置"菜单的实现…………… 128
5.3.1　增加记录…………………… 79	7.7　怪物敌人功能的实现………………… 132
5.3.2　修改与删除记录…………… 79	7.8　怪物BOSS功能的实现……………… 134
5.3.3　自定义数据类型与字节数组的转换技巧…………………………… 80	7.9　人物魔法技能功能的实现…………… 136
	7.10　游戏碰撞检测功能的实现………… 137
5.3.4　利用RMS实现对象序列化… 81	7.11　游戏按键检测功能的实现………… 139
5.4　RecordStore的高级操作…………… 82	7.12　游戏主要逻辑循环功能的实现…… 142
5.4.1　RecordEnumeration遍历接口… 82	7.13　其他功能的实现…………………… 150
5.4.2　RecordFilter过滤接口…… 84	7.13.1　游戏加载进度条类…… 150
5.4.3　RecordComparator比较接口… 85	7.13.2　游戏道具类…………… 152
5.4.4　RecordListener监听器接口… 86	7.13.3　游戏公共参数资源配置的实现…… 153

7.14 游戏实现效果图 ················· 154
7.15 本章小结 ····················· 155
第 8 章 网络编程 ····················· 156
　8.1 移动网络编程概述 ············· 156
　　8.1.1 CLDC 通用连接框架 ········· 156
　　8.1.2 CLDC 通用连接类 ··········· 157
　8.2 HTTP 编程 ··················· 160
　　8.2.1 MIDLet 连接到 HTTP 服务器上 ······ 160
　　8.2.2 获取 HTTP 连接的基本信息 ········· 161
　　8.2.3 手机客户端与 HTTP 服务器通信 ····· 163
　8.3 Socket 套接字编程 ············· 176
　　8.3.1 客户端与服务器的套接字连接 ······· 176
　　8.3.2 套接字连接可以得到的基本信息 ····· 177
　　8.3.3 套接字连接通信 ··············· 179
　8.4 UDP 数据报编程 ··············· 187
　　8.4.1 客户端与服务器端数据报连接 ······· 187
　　8.4.2 数据包的传递 ················· 188
　8.5 本章小结 ····················· 196
第 9 章 MMAPI 多媒体程序设计 ······ 197
　9.1 移动媒体 API（MMAPI）概述 ········· 197
　　9.1.1 MMAPI 的体系结构 ············ 197
　　9.1.2 管理器 Manager 类 ············· 198
　　9.1.3 播放器 Player 接口 ············ 199
　　9.1.4 数据源 DataSource 类 ·········· 201

　　9.1.5 控制器 Control 接口 ············ 201
　9.2 音频播放 ····················· 201
　9.3 视频播放 ····················· 206
　9.4 手机拍照的实现 ··············· 212
　9.5 本章小结 ····················· 219
第 10 章 无线消息程序设计 ··········· 220
　10.1 无线消息概述 ················ 220
　　10.1.1 GSM 短消息服务 ············ 220
　　10.1.2 GSM 小区广播 ·············· 221
　10.2 WMA 概述 ·················· 222
　10.3 使用 WTK 中的 WMA 控制台 ········ 223
　　10.3.1 配置和启动 WTK 中的 WMA 控制台 ······················· 223
　　10.3.2 使用 WMA 控制台发送文本消息·· 225
　　10.3.3 使用 WMA 控制台发送小区广播·· 227
　　10.3.4 使用 WMA 控制台发送多媒体消息 ························· 228
　10.4 编写利用 WMA 控制台收发短消息的程序 ························· 230
　　10.4.1 发送和接收 SMS 消息 ········· 230
　　10.4.2 发送和接收二进制消息 ········ 236
　　10.4.3 发送和接收多媒体消息 ········ 243
　10.5 本章小结 ···················· 251

第 1 章 Java ME 概述

本章主要介绍 Java ME 的相关背景知识。读者需要掌握以下知识点：
- Java ME 的 3 层体系结构。
- CLDC 类库。
- MIDP 2.0 的新特性。
- CLDC/MIDP 总体体系结构。

1.1 Java ME 体系结构

为了适应移动数据的发展，推进无线电子商务等业务的发展，J2ME（Java 2 Micro Edition）即用于嵌入式系统的 Java 被引入无线领域（又称 Java ME）。Java ME 的出现实际上是 Java 技术的回归。作为 Java 2 平台的一部分，Java ME 与 J2SE、J2EE 一起，为无线应用的客户端和服务器端建立了完整的开发和部署环境。随着 Java ME 的应用，它为移动互联引入了一种新的模型，即允许手机从互联网上下载各种应用程序，并在手机上创造可执行环境离线运行这些程序。由于定义了可执行程序下载的标准，并在手机上创立了可执行环境和程序开发语言，由此，在移动通信业第一次为软件开发商创造了巨大的商机，手机用户在得到丰富应用体验的同时，也大大提高了运营商的网络流量。

1.1.1 Java 的版本

Java 在 10 多年的发展历程中，已经成长为一个全面而成熟的面向对象应用程序开发平台，它适用于广泛的、异构的编程环境，这些应用的涉及面非常广，从企业级的服务器应用到传统的桌面应用以及各式各样面向小型设备的嵌入式应用。

Java 2 平台包括 3 个版本，每个版本都针对不同的用户群，如图 1-1 所示。这 3 个版本具体为：

- Java 2 平台企业版（J2EE）：用于企业级分布式系统的设计与开发。
- Java 2 平台标准版（J2SE）：用于传统 PC 桌面应用程序开发。
- Java 2 平台微型版（J2ME）：主要面向消费电子产品和嵌入式设备的软件开发。

说明：Java SDK 1.2 及以后的版本都统一改名为 Java 2，因此这些名字中都带有 2。

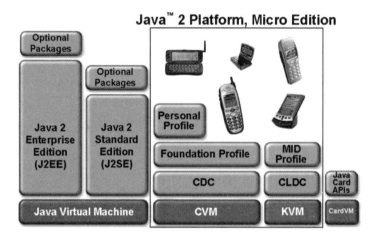

图 1-1　Java 的版本结构

Java 2 面向市场的每一个版本都有其自己的虚拟机，这些虚拟机都为其目标应用做了特别优化。

1.1.2　Java ME 的 3 层体系结构

Java ME 用于为信息家电市场提供应用服务，这些信息家电包括传呼机、移动电话、像 Palm 这样的个人数字助手（PDA）、电视机顶盒、POS 终端以及其他的消费电子设备，而且每一种家电设备又有不同的特性和界面。

为了满足消费者和嵌入式市场不断发展和多样化的需求，Java ME 体系结构采用模块化、可扩展的设计。这种设计是通过一个 3 层软件模型来实现的，该模型构建于本地操作系统之上。

Java ME 的 3 层体系结构依照各种设备的资源特性，将 Java ME 技术架构分为简表（Profile）、配置（Configuration）和 Java Virtual Machine（JVM）3 层，然后再进一步细分，这使 Java ME 能够在每一类设备的限制下工作，同时能够提供最低限度的 Java 语言功能性，如图 1-2 所示。

第 1 章 Java ME 概述

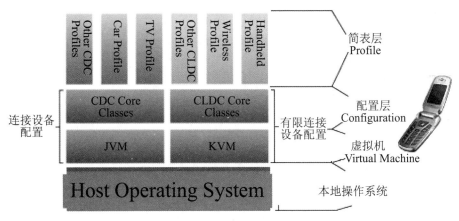

图 1-2　Java ME 的分层结构

- Java 虚拟机（JVM）层：这一层是针对设备本地操作系统定制的 Java 虚拟机的实现，支持特定的 Java ME 配置，就像使用所有 Java 技术一样，Java ME 的核心也在一种虚拟机中。
- 配置（Configuration）层：面对的是大量各种不同的小型嵌入式设备，它们在外观和功能上均各不相同。Java ME 对这些设备进行分类，将一些共性提取出来形成适合于某个范畴中设备可用的规范，称为"配置"。读者也可以将配置理解为对硬件的描述，所以通过定义配置的方法就能够清楚地描述硬件功能。
- 简表（Profiles）层：简表层定义了特定系列设备上可用的应用程序编程接口（API）的最小集。简表在一个特定的配置上面实现。应用程序是针对特定的简表编写的，因此可以移植到支持该简表的任何设备上。另外，一个设备可以同时支持多个简表。用户和开发人员看到最多的就是这一层。

Java 虚拟机是 Java ME 技术的核心，配置和简表则提供特殊环境的类应用程序接口。配置是用于一组通用设备的最小的 Java 平台，而简表则为具体的设备家族或特别的应用程序提供更具体的功能。

Java 虚拟机和构建于此虚拟机之上的配置规范一起代表了某一类设备的基本能力，而更进一步的设备分类上的区别是通过简表层提供的 API 实现的。为了满足更多新的应用的需要，简表可以通过扩充类库来强大自己的功能。

1.1.3　虚拟机（KVM）

KVM 是一个专门为小型、资源受限设备所设计的紧凑的、便携的 Java 虚拟机，是用于 Java ME 平台最小的虚拟机，并且是用于 CLDC 配置的虚拟机。Java ME 应用程序并不一定非要使用 KVM，Java ME 技术支持使用任何虚拟机，不过至少应当有 KVM 这样的功能。

KVM 是完全从头开始编写的，其设计目标包括：
- 虚拟机的大小和类库为 60KB～80KB 左右。
- 内存占用为几十 KB。
- 在具有 16 位和 32 位处理器的设备上，有相当的性能。
- 高度可移植和可扩展，特定于机器的代码总量很少。
- 多线程和垃圾回收是独立于系统的。
- 可以对虚拟机的组件进行配置，以适合于特定设备，从而增强了灵活性。

为什么称为 KVM 呢？K 代表 Kilo。这样命名是因为它的内存容量是用几十 KB 来衡量的。其典型地应用于移动电话、传呼机、个人信息管理器和便携式终端等。

说明： 由于这一历史原因，造成了很多名词上的困扰。许多早期的文章把 KVM 与 com.sun.kjava 包合称为 KVM，来共同表示这一技术。而现在所说的 KVM 单指 Sun 的 CLDC 虚拟机引用实现。甚至还有人用 K-Java 作为 Java ME 的代称，表示基于 KVM 的 Java，这很容易和正式的 Java ME 标准以及早期的 com.sun.kjava 包相混淆。读者在看这类文章时需要注意文章的时间和背景，并采用严格的正式名称。

KVM 实现所需的最小内存空间大约是 128KB，包括虚拟机、最小的库和运行 Java 应用所需要的堆空间。一个更加典型的实现需要总共 256KB 内存空间，其中 32KB 作为应用运行时的堆空间，60KB～80KB 用于虚拟机本身，剩余的为类保留。

1.2 有限连接设备配置表（CLDC）

CLDC 提供了一个适合于小型、资源受限、连接受限设备上使用的标准 Java 平台。这种设备有移动电话、PDA 等。为此，它在每个方面都进行了大量的优化。它的虚拟机很小，并且不支持 Java 语言的一些特性。

1.2.1 CLDC 概览

CLDC 起源可以追溯到 1999 年 JavaOne 大会上介绍的 Sun 的第一个袖珍版 Java 和第一个 KVM 以及相关的类库，虽然 CLDC 和所有的配置都满足成为虚拟机的条件，但是它本身还不是虚拟机，CLDC 的引用实现只是包含在当前分布中的 KVM。

根据规范，运行 CLDC 的设备应该有：
- 至少 192KB 的内存空间。
- 16 或 32 位处理器。
- 一个有限的电源供给（通常使用电池）。
- 有限的或间断的网络连接性（9600 bps 或更少）。
- 多样化的用户界面甚至没有用户界面。

说明：由于增加了浮点运算等功能，CLDC 的内存需求从 1.0 的 160KB 上升到 1.1 的 192KB。其中 160KB 用于存储实际的虚拟机和本身的类库，CLDC 规范假定应用程序能够在 32KB 这样小的 Java 堆栈空间中运行。

通常说来，这个配置是为个人化的、移动的、有限连接信息设备而设计的，如传呼机、移动电话和 PDA 等。与 J2SE 相比，CLDC 缺少下列特征：

- AWT（抽象窗口开发包）、Swing 或其他图形库。
- 用户自定义类装载器。
- 类实例的最终化。
- 弱的引用（在 CLDC1.1 中已经引入）。
- RMI。
- Reflection（映射）。

如果要理解为什么 CLDC 去除这么多 J2SE 中重要的类和特征，可以回顾一下与 CLDC 相关的两条基本原理。首先，它只有 512KB 的内存空间，而 RMI 和映射需要的内存太大。其次，配置必须满足为一组通用设备提供最小的 Java 平台。在个人移动信息设备领域中，许多系统都不能支持 J2SE 中的众多高级特征。例如，许多系统没有或不提供访问一个文件系统的功能或权限，因此与文件有关的类也被丢弃了。而且错误处理是一个代价非常高的过程处理，在许多消费电子设备中，故障恢复是很难的甚至是不可能的。所以在 CLDC 中，许多错误处理类也被删除了。

1.2.2 CLDC 中使用的 J2SE 类

CLDC 包含一个很小的 J2SE 子集，因为其他的类对虚拟机和本地运行环境的依赖性都比较大，而且会消耗大量的资源，因此被忽略掉了。下面介绍 CLDC 支持的类。

1. java.lang 中的类

（1）系统类

J2SE 类库中包含了几个同 Java 虚拟机密切相关的类，经常使用的几个 Java 工具需要有几个类的支持，如标准的 Java 编译器（javac）要求 String 和 StringBuffer 中的特定方法是可用的。所以 CLDC 中支持的系统类主要包括以下几个：

- java.lang.Object
- java.lang.Class
- java.lang.Runtime
- java.lang.System
- java.lang.Thread
- java.lang.Runnable（接口）
- java.lang.String

- java.lang.StringBuffer
- java.lang.Throwable

（2）数据类型类

CLDC 中同样支持 J2SE 中的基本数据类型，这些类都是 J2SE 相应类型的子集。它们是：

- java.lang.Boolean
- java.lang.Byte
- java.lang.Short
- java.lang.Integer
- java.lang.Long
- java.lang.Float（从 1.1 版本开始支持）
- java.lang.Double（从 1.1 版本开始支持）
- java.lang.Character

（3）错误类

- java.lang.Error
- java.lang.NoClassDeFoundError（从 1.1 版本开始支持）
- java.lang.OutOfMemoryError
- java.lang.VirtualMachineError

（4）异常类

- java.lang.Exception
- java.lang.ArithmeticException
- java.lang.ArrayIndexOutOfBoundsException
- java.lang.ArrayStoreException
- java.lang.ClassCastException
- java.lang.ClassNotFoundException
- java.lang.IllegalAccessException
- java.lang.IllegalArgumentException
- java.lang.IllegalMonitorStateException
- java.lang.IllegalThreadStateException
- java.lang.IndexOutOfBoundsException
- java.lang.InstantiationException
- java.lang.InterruptedException
- java.lang.NegativeArraySizeException
- java.lang.NullPointerException
- java.lang.RuntimeException
- java.lang.NumberFormatException

- java.lang.SecurityException
- java.lang.StringIndexOutOfBoundsException

（5）弱引用

- java.lang.ref.Reference（从 1.1 版本开始支持）
- java.lang.ref.WeakReference（从 1.1 版本开始支持）

2．java.io 中的类

（1）输入/输出类

- java.io.InputStream
- java.io.OutputStream
- java.io.ByteArrayInputStream
- java.io.ByteArrayOutputStream
- java.io.DataInput（接口）
- java.io.DataOutput（接口）
- java.io.DataOutputStream
- java.io.DataInputStream
- java.io.Reader
- java.io.Writer
- java.io.InputStreamReader
- java.io.OutputStreamReader
- java.io.PrintStream

（2）在 java.io 中还有几个相对于输入/输出处理的异常类

- java.io.EOFException
- java.io.InterruptedIOException
- java.io.IOException
- java.io.UnsupportedEncodingException
- java.io.UTFDataFormatException

3．java.util 中的类

（1）集合容器类

- java.util.Vector
- java.util.Stack
- java.util.Hashable
- java.util.Enumeration（接口）

（2）日历和时间类

CLDC 还包含了 J2SE 中相对于时间和日期的类，例如 java.util.Calendar、java.util.Date 和 java.util.TimeZone 类的子集。为了节省空间和其他硬件资源，CLDC 规范中仅要求电子设备支

持一个时区,是否支持其他时区以及具体支持什么时区可以由硬件设备制造商来决定。以下为 3 个日历和时间类:

- java.util.Calendar
- java.util.Date
- java.util.TimeZone

(3) 异常类

CLDC 中的异常除了 java.lang 和 java.io 两个包之外,还包括 java.util 的异常。

- java.util.EmptyStackException
- java.util.EmptyNoSuchElementException

4. 其他附加类

除了上面所描述的几个集合的类外,CLDC 中还提供了两个附加类。其中,java.util.Random 是一个简单的随机数生成类,可以由比较简单的算法生成随机数,而 java.lang.Math 是一个数学函数库的类,其提供的方法如 min、max 等可以用于基本的数学函数运算。在 CLDC 1.1 中,java.lang.Math 类还支持浮点三角函数和平方根运算,并增加了其他一些工具函数,如 ceil 和 floor。

- java.util.Random
- java.lang.Math

1.2.3　CLDC 专用类

　　J2SE 和 J2EE 类库通过 java.io 和 java.net 包提供的一系列功能可用来处理访问存储和网络系统的输入和输出。尽管如此,保证所有这些功能都适合只有几百 KB 存储空间的小型设备是很困难的。于是 CLDC 针对 J2SE 网络和 I/O 类进行抽象、泛化,这些新系统的一个目标就是 J2SE 类的严格子集。这样才能容易地映射到通用的底层硬件或者任何 J2SE 实现,而且保证具有较好的扩展性、灵活性和一致性,以支持新的设备和协议。

　　一般来说,资源受限设备在对网络和存储类库的要求上差异非常明显。例如,需要处理包交换网络的设备制造商通常会需要基于数据报的通信机制,而需要处理线路交换网络的设备制造商则需要基于流的连接。考虑到严格的内存限制,如果制造商支持了特定类型的网络功能,通常都不希望再支持其他机制。所有这些都使得设计 CLDC 的网络设施非常具有挑战性,特别是由于 Java ME 配置不允许定义可选的特性。而且,如果存在多种网络机制和协议,则可能会使应用程序开发者非常困扰,特别是当他们不得不自己处理底层协议时。

　　于是,CLDC 采用"通用连接框架 (GCF)",取代了 J2SE 中的网络 API,为 CLDC 提供网络连接功能。所有 Java ME/CLDC 应用程序都有权访问 HTTPS 功能,但在 MIDP 1.0 规范中并不正式需要 HTTPS 支持。考虑到 HTTPS 在移动商业中显而易见的重要性,许多 MIDP 设备供应商已经将对 HTTPS 的支持添加到它们自己的 MIDP 运行时实现中。Sun Microsystems 也在其 Java ME Wireless Toolkit 版本 1.0.2 及后续版本中添加了 HTTPS 支持。在 MIDP 2.0 规范中，HTTPS 支持将成为正式需求。CLDC 1.0 规范（CLDCSPEC）为 MIDP 网络定义了一个抽象的通用连接框架。该框架定义了各种网络 API 的接口格式，但没有对任何网络协议作具体实现。

该框架支持的功能包括：
- 各种基于流的连接。
- 各种数据报。

按照 MIDP 1.0 规范（MIDPSPEC），基于 CLDC 通用连接框架的 MIDP 实现至少需要实现 HTTP1.1 上（RFC2616）基于流的各种客户端连接。一个 MIDP 实现也可以实现 CLDC 通用连接框架中其他各种网络 API（例如 TCP/IP 套接字、数据报等）。但是，因为 MIDP 1.0 规范仅将 HTTP 网络协议定义为基本配置，所以为了 MIDlets 的完全可移植性，不应使用 MIDP 规范中除 HTTP 网络 API 之外的其他 API。

1.2.4 CLDC 1.1 的新特性

CLDC 1.1（JSR139）专家组成员对 CLDC Specification 1.0 版基本满意，他们认为在新的规范中不需要做什么根本上的修改。因此，CLDC Specification 1.1 版基本上只是一个增补版，并且是对 CLDC Specification 1.0 版完全向后兼容的。一些重要的新功能，如对浮点的支持，被加入到这个新版本中。

CLDC 1.1 相对于 CLDC 1.0（JSR30）版本本质上没有变化，只是随着硬件水平的不断提高，CLDC 1.1 在兼容性和可用性上做了如下一些改进：
- 增加了对浮点数据的支持。
- CLDC 1.1 支持所有跟浮点运算相关的字节码指令。
- 由于允许使用浮点运算，设备的最小内存从 160KB 提高到 192KB。
- 加入了 Float 和 Double 两个类，其他类库中的类也增加了一些方法以处理浮点数。
- 增加了对弱引用（Weak Reference）的支持（只是 J2SE 的弱引用类的一个小子集）。
- 更加明确了对错误处理的要求，增加了一个新的错误类 NoClassDefFoundError。
- Thread 对象同 J2SE 中的线程一样被命名。
- 引入了 Thread.getName()方法来获知线程的名字，并且 Thread 类从 J2SE 继承了几个新的构造函数。
- 线程可以被 Thread.interrupt()方法中断。
- Calendar、Date 和 TimeZone 类被重新设计。
- 增加了一份更详细的验证器规范作为 CLDC Specification 1.1 版的附录（《CLDC Byte Code Typechecker Specification》）。

类库中还加入了以下方法用于处理浮点数据：

- java.lang.Integer.doubleValue()
- java.lang.Integer.floatValue()
- java.lang.Long.doubleValue()
- java.lang.Long.floatValue()
- java.lang.Math.abs(double a)
- java.lang.Math.abs(float a)
- java.lang.Math.max(double a, double b)
- java.lang.Math.max(float a, float b)
- java.lang.Math.min(double a, double b)
- java.lang.Math.min(float a, float b)
- java.lang.Math.ceil(double a)
- java.lang.Math.floor(double a)
- java.lang.Math.sin(double a)
- java.lang.Math.cos(double a)
- java.lang.Math.tan(double a)
- java.lang.Math.sqrt(double a)
- java.lang.Math.toDegrees(double angrad)
- java.lang.Math.toRadians(double angrad)
- java.lang.String.valueOf(double d)
- java.lang.String.valueOf(float f)
- java.lang.StringBuffer.append(double d)
- java.lang.StringBuffer.append(float f)
- java.lang.StringBuffer.insert(int offset, double d)
- java.lang.StringBuffer.insert(int offset, float f)
- java.io.DataInput.readDouble()
- java.io.DataInput.readFloat()
- java.io.DataInputStream.readDouble()
- java.io.DataInputStream.readFloat()
- java.io.DataOutput.writeDouble(double v)
- java.io.DataOutput.writeFloat(float v)
- java.io.DataOutputStream.writeDouble(double v)
- java.io.DataOutputStream.writeFloat(float f)
- java.io.PrintStream.print(double d)
- java.io.PrintStream.print(float f)
- java.io.PrintStream.println(double d)

- java.io.PrintStream.println(float f)
- java.util.Random.nextDouble()
- java.util.Random.nextFloat()

增加了以下成员变量以支持浮点数运算：
- java.lang.Math.E
- java.lang.Math.PI

增加了 java.lang.ref.Reference 和 java.lang.ref.WeakReference 类。

增加了一个错误类 NoClassDefFoundError。

重新设计了 java.util.Calendar、java.util.Date 和 java.util.TimeZone 类，类中增加了许多成员变量和方法以便于 J2SE 更加一致。

Thread 类中增加了以下新的方法和构造函数：
- Thread.getName()
- Thread.interrupt()
- Thread(Runnable Target, String name)
- Thread(String name)

另外还加入了以下成员变量和方法：
- java.lang.Boolean.TRUE 和 java.lang.Boolean.FALSE
- java.lang.String.intern()
- java.lang.String.equalsIgnoreCase()
- java.util.Date.toString()
- java.util.Random.nextInt(int n)

1.3 MIDP

MIDP（Mobile Information Device Profile）定义了针对移动信息设备（主要指智能手机和一部分具有无线通信功能的 PDA）的图形界面、输入和时间处理、持久性存储、无线电话网络连接之上的一些消息处理（例如短消息）、安全等 API，并且考虑到了移动信息设备的屏幕和内存限制。类似于 J2SE 中的 Applet 框架，MIDP 提供了基于 javax.microedition.midlet 包的 MIDlet 应用程序框架。

1.3.1 设备需求

1. 硬件需求

MIDP 专家组（MIDPEG）的目标是建立一个开放的用于移动信息设备（MID）的第三方应用开发环境。为了达到这个目标，MIDPEG 定义了 MID 应该满足的最小特征。

显示屏：
- 屏幕尺寸：96×54
- 显示色彩：1-bit
- 像素形状：1:1

输入方式：

单手输入的键盘、双手输入的键盘（传统的 QWERTY 键盘）、触摸屏。

内存：
- 128KB 非易失性存储器(non-volatile memory)，用于 MIDP 组件（MIDP 2.0 需要 256KB 稳定存储）。
- 8KB 非易失性存储器，用于应用创建所需的固有数据。
- 32KB 可变存储内存，用于 Java 运行环境（例如 Java 堆栈）（MIDP 2.0 需要 128KB 内存）。

网络支持：
- 双向的、无线的、间断的、有限带宽的连接。

声音：
- 能够通过硬件和软件算法演奏声音（MIDP 2.0）。

说明：这里所指出的存储需求只是 MIDP 组件的存储需求，不包括 CLDC 和其他系统软件的存储需求。非易失（Non-volatile）的意思是指当用户在关闭设备和打开设备期间，仍然能保持其内容不丢失。这里通常认为对稳定存储的操作是只读模式的，进行写操作需要特殊步骤。稳定存储的例子包括只读存储器（ROM）、闪存（Flash）和有后备电池的 SDRAM。易失（Volatile）的意思是指在用户关闭设备和打开设备期间不保存它的内容，通常认为对它的操作模式包括读模式和写模式，对它的访问不需要特殊的步骤。易失存储最常见的类型是 DRAM。

2．软件需求

对于具有前面所提到的硬件特征的设备，它们的软件能力也有很大发展空间。和用户的桌面计算机模型不同，后者具有大型的、垄断性的系统软件体系结构，而 MID 范围内有很多种系统软件，而且各有不同。例如，有的操作系统功能比较完全，支持多线程和层级文件系统；而另一些可能只有很小的、基于线程的操作系统，没有文件系统的概念。面对如此多样的 MID 操作系统，MIDP 规范对软件需求做出如下最小假设：

- 管理底层硬件的最小内核（处理中断、异常和小型调度等）必须提供至少一个可调度实体来运行 Java 虚拟机（JVM），它不必支持单独的寻址空间（或进程），也不必保证实时调度或者延迟行为。
- 具有从稳定存储中读写的机制，以支持持久存储操作。
- 对设备无线网络的访问（读、写），以支持联网操作。
- 提供基于时间操作的基本机制，支持用于持久存储中记录的时间戳和 timer 组件的基本操作。
- 提供写点阵图像显示屏的最小能力。
- 提供 3 种输入方式（参看硬件需求）中的一种或几种捕获用户输入的机制。
- 提供一种管理设备上应用程序生命周期的机制。

1.3.2　MIDP 的总体体系结构

图 1-3 列出了用于无线应用的 Java ME 高层体系结构。这一结构可分为 5 层，从下向上依次为：

- MID 硬件层（MID Hardware Layer）：主要包括手机或者双向寻呼机。
- 本地系统软件层（Native System Software Layer）：包括由设备生产商提供的本地操作系统和系统库。
- KVM 层（KVM Layer）：为 Java 无线应用提供了实时运行环境。
- CLDC 层（CLDC Layer）：为 Java 无线应用提供了核心 Java API。
- MIDP 层（MIDP Layer）：提供了 GUI（图形用户界面）、持续存储、网络等类库。

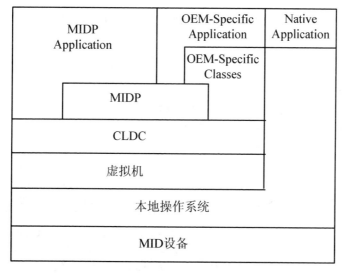

图 1-3　MIDP 体系结构

CLDC 和 MIDP 标准化成果的最高目标是建立一个具有高移植性、安全性、资源占用少的应用开发平台，使第三方可以为这些资源受限的互联设备进行开发。CLDC 作为一个通用的底层标准平台，定位于所有类型的资源受限设备，而 MIDP 则是建立于此之上的特定的用于无线双向通信设备的简表。

1.3.3 MIDP 类库

MIDP 类库建立在 CLDC 类库基础之上，因此许多类都来自 J2SE。为了实现特定功能，MIDP 增加了一些新的类库，这些类总结如表 1-1 所示。

表 1-1 MIDP 中的类包总结

核心包	包描述
java.io	通过数据流提供系统的输入输出
java.lang	MIDP 语言包，在 CLDC 中 java.lang 包的基础上增加了类（来自 J2SE）java.lang.IllegalStateException。当有非法的转换请求时抛出这个异常，例如在一个 TimerTask 安排中调用或者在用户界面组件容器中请求
java.util	MIDP 工具类，在 CLDC 中 java.util 包的基础之上增加了以下类（来自 J2SE）： java.util.Timer java.util.TimerTask Timer 用于安排一个后台线程使它在以后执行，TimerTask 通过使用 Timer 安排一个单次执行任务，或者以一定间隔重复执行的任务
应用程序生命周期 javax.microediton.midlet	midlet 包定义了 MIDP 应用程序以及应用程序和它所运行环境之间的交互
用户界面 javax.microedition.lcdui javax.microedition.lcdui.game	为 MIDP 应用程序提供用户界面 API 为 MIDP 应用程序提供游戏开发的高级 API（MIDP 2.0 新增）
持久存储 javax.microedition.rms	这个包用来为 MIDlet 提供持久存储的机制，应用程序可以存储数据，在以后需要的时候获取这些数据
网络 javax.microedition.io	MIDP 提供了基于 CLDC 通用连接框架的网络支持，在 CLDC 包的基础之上，增加了类 javax.microedition.io.HttpConnection，这个类用于建立 HTTP 连接。在 MIDP 2.0 中除了支持 HTTP 连接，还支持 HTTPS、报文、Socket 通信以及串口通信。另外，MIDP 2.0 还支持服务器 Push 体系架构
公开密钥 javax.microedition.pki	提供用于鉴别安全网络信息的数字认证（MIDP 2.0 新增）
声音媒体 javax.microedition.media javax.microedition.media.control	该包为 MIDP 应用程序提供音、视频等多媒体功能，该包遵循 JSR-135 规范（MIDP 2.0 新增） 该包为播放器提供了一些特定的控制功能（MIDP 2.0 新增）

一些功能并不在 MIDP 的定义范围内，这些功能包括：

- 系统层的 APIs：MIDP APIs 的重点是针对应用的开发者，而不是系统开发。因此一些底层的涉及系统接口的 APIs，如无线信息设备的电源管理、语音的编码模块就属于系统层的 APIs 而不在 MIDP 所讨论的范围内。

- 底层的安全：MIDP 规范不涉及底层的安全功能，它只使用 CLDC 提供的底层的安全机制。

1.3.4　MIDP 2.0 的新特性

MIDP 2.0 是由 JCP 制订的，由大约 50 个公司共同参与设计。MIDP 2.0 规范设计的目的是定义一个新体系架构以及相应的API，从而为第三方的移动信息设备（MIDs）应用的开发提供一个开放的标准环境。

MIDP 2.0 规范是在 MIDP 1.0 规范的基础上设计的，它保证了同 MIDP 1.0 的兼容性。即在 MIDP 1.0 上编写的 MIDlets 能在 MIDP 2.0 上执行。

根据 Java ME 定义的体系架构，MIDP 被设计在 CLDC 的基础上运行。虽然 MIDP 2.0 规范是在 CLDC 1.0 所提供的功能的基础上制订的，但它仍能运行在 CLDC 1.1 之上以及之后的更新版本。但 MIDP 2.0 的实现最好在 CLDC 1.1 的基础上。

为了支持对声音的处理（例如播放 WAV 文件），MIDP 2.0 增加了一个可选包 ABB（Audio Building Block）。在以前，ABB 只包含在 Mobile Media API（MMAPI）中，现在将其纳入到 MIDP 中来，开发者就可以不依赖 MMAPI。当然，如果使用了 MMAPI，可以为移动设备开发更多的功能，例如可以在 PDA 上播放视频流等。

对 Java ME 游戏开发的支持可能是开发者和用户都企盼已久的。MIDP 2.0 提供的游戏 API 使得游戏本身可以更充分地利用设备自身的图形处理功能。它的出现无疑大大简化了 Java ME 游戏的开发工作，同时也使得开发者可以更多地控制程序的图形处理性能。

在通信方面，MIDP 1.0 仅支持 HTTP，MIDP 2.0 则增加了对 HTTPS、报文、Socket 通信以及串口通信的支持。另外，MIDP 2.0 还支持服务器 Push 体系架构，这样手机能够收到来自服务器的报警、消息或者广播，根据要求启动手机上的应用程序进行操作。

支持 Over-the-air（OTA）Provisioning 是 MIDP 2.0 的一个重要新特点，它使得用户能够动态地部署和更新移动设备上的应用程序。新版的 MIDP 规范中规定了如何在移动设备上发现、安装、更新和删除 MIDlet 套件；同时，提供应用程序下载的服务提供商还能够判断该 MIDLET 套件是否能够运行在申请下载的设备上，并且从设备上获取安装、更新和删除的信息。MIDP OTA Provisioning 模型为移动服务提供商提供了单一、标准的部署 MIDP 应用程序的途径。该模型已经被众多技术领先的移动设备制造商和服务提供商所采纳。

自从人类步入网络时代以后，网络安全始终是一个很关键的问题。MIDP 2.0 增加了强大的"端到端"的安全模型。一方面，MIDP 2.0 支持 HTTPS，可以对传输的数据进行加密；另一方面，MIDP 2.0 采用了和 MIDP 1.0 不同的安全机制，采用安全域来确保未经授权的 MIDlet 套件无法访问受权限控制的数据、应用程序以及其他网络和设备资源。

无线信息设备（MIDs）的功能多种多样。但 MIDP 规范并不是要对这些设备的所有功能都要加以定义，并提供 APIs 编程。相反 MIDP 1.0 和 MIDP 2.0 的专家组同意只针对有通用并且成功实施的功能的需求制订相应的 APIs。在 MIDP 1.0 中这些功能包括：
- 应用的下载。
- 应用的生命周期。
- 端到端的传输（http）。
- 网络连接。
- 数据库存储。
- 计时器。
- 用户界面。

通过用户对 MIDP 1.0 的使用经验和反馈，MIDP 2.0 的专家组在 MIDP 1.0 APIs 的基础上又新加了下面的 API：
- 应用的下载和计费。
- 端到端的安全传输（https）。
- 应用的数字签名和域的安全模式。
- MIDlet 的 push 注册（server push model）。
- 声音。
- 游戏开发。

1.3.5　MIDP 2.0 的安全机制

MIDP 1.0 规范在制订时由于当时的无线网络尚处于发展阶段，因此在考虑 MIDP 1.0 的安全机制时没有太多考虑应用在网络传输过程中的安全性设置，而是把重点放在无线应用在设备上运行时的安全性上。因此 MIDP 1.0 规范通过沙箱的安全机制来提供 MIDlet 套件运行时的安全性，该机制保证 MIDlet 所能调用的 APIs 不能访问设备的敏感信息和功能。在 MIDP 2.0 安全机制的制订过程中，专家组采纳了用户对 MIDP 1.0 安全机制所提出的意见：即 1.0 版本的安全性太"强"而使得很多无线设备的功能无法完全使用。

在 MIDP 2.0 中的敏感操作基本上都是和网络连接相联系的，例如通过 HTTP 协议联网。这些操作可能会让用户付费甚至存在安全隐患。如果要把这些敏感功能的 APIs 开放而让用户的应用访问，就必须引入一种新的安全机制。该机制就是 Java 1.2/2 的域安全机制。在 MIDP 2.0 中对不被信任的 MIDlet 只能运行于该沙箱安全机制中。任何一种 MIDP 2.0 的实现，必须支持非信任的 MIDlet 在 MIDs 设备上的运行。

在 MIDP 2.0 的域安全机制中，主要涉及以下概念：
- 许可：用来保护对敏感 API 的访问。应用程序通过对敏感 API 提出许可申请来试图获得相应的权限。
- 保护域：是 MIDlet 套件所允许的权限访问的集合。
- 权限访问：是指必须通过授权才能使用的 APIs 或功能。

- 信任 MIDlet 套件：指该 MIDlet 套件能通过验证并且 JAR 文件的完整性能被保证，并能被该设备上的某一保护域所信任。
- 配置文件：一个配置文件中包含许多域或别名的定义。每个域由赋予的权限和用户操作所构成。

当进行某些敏感操作时，例如访问网络，系统会查询用户。如果用户不许可，系统就会抛出 SecurityException。下面列出了在 MIDP 2.0 中定义的许可：

- java.microedition.io.Connector.http
- java.microedition.io.Connector.socket
- java.microedition.io.Connector.https
- java.microedition.io.Connector.ssl
- java.microedition.io.Connector.datagram
- java.microedition.io.Connector.serversocket
- java.microedition.io.Connector.datagramreceiver
- java.microedition.io.Connector.comm
- java.microedition.io.PushRegistry

说明：虽然这些许可的名称和类名很类似，但是两者并非等同，不可混淆。

与许可相关联的一个概念是保护域。保护域就是一组许可以及作用在这组许可上的交互模式。一个设备上有多个保护域，MIDlet 都是运行在不同的保护域中的。不同的设备提供的保护域可能是不同的。

不过，MIDP 规范定义了非可信域的推荐范围。原则上，非可信域提供较少的许可，并且这些许可的确认要经过用户操作来完成。规范还定义了非可信域的用户交互模式（授权模式）：blanket（总是允许访问）、session（下次不再询问）、oneshot（每次询问）。不过不是每个设备都开放这 3 种交互模式给非可信域的每个敏感操作。例如，如果某些非可信域的交互模式只提供 oneshot（每次询问），这意味着运行在这个域的 MIDlet 每次进行这个操作时，都要获得用户确认。其他 MIDP 规范未提到的非可信域则由设备决定。

MIDP 2.0 引入了新概念——信任的 MIDlet。对于信任的 MIDlet，它能使用一些敏感的和被限制使用的 APIs。当无线信息设备检测到一个 MIDlet 套件是可信任时，则它对 APIs 的访问由相应的安全域的策略定义。将在后面的章节中详细介绍。当验证一个 MIDlet 套件是否能信任时发生错误，则该 MIDlet 套件必须被拒绝执行。

1. 非信任的 MIDlet 套件

一个非信任的 MIDlet 套件是指 MIDlet 套件的下载源和 JAR 文件的完整性不能被下载的设备所信任。对于非信任的 MIDlet 套件必须只能运行于非信任的安全域中，在该域中，对受保护的 APIs 或功能不能访问或由设备使用者确定可以访问或不可以访问。任何一个与 MIDP 1.0 兼容的 MIDlet 套件能在 MIDP 2.0 的实现上用非信任的模式运行。对于非信任的 MIDlet 套件，它能访问它的 JAR manifest 和 application descriptor 文件以及下面的 APIs。

- javax.microedition.rms
- javax.microedition.midlet
- javax.microedition.lcdui
- javax.microedition.lcdui.game
- javax.microedition.media
- javax.microedition.media.control

非信任安全域对以下 APIs 或功能的访问必须由用户确定可以或不可以（申请许可）。

- javax.microedition.io.HttpConnection
- javax.microedition.io.HttpsConnection

2. 可信任的 MIDlet 套件安全

可信任的 MIDlet 套件安全是建立在保护域的机制上。每一保护域定义了该域中的 MIDlet 套件的访问权限，该保护域的定义者可以定义设备怎样确认和验证它能信任的 MIDlet 套件，并把该域所允许和授权使用的受保护的 APIs 和功能绑定给该 MIDlet 套件。设备所使用的验证和信任 MIDlet 套件的机制的定义是分开的，这样有利于人们根据设备、网络和商业模式的不同来进行选择。可信任的 MIDlet 套件使用 X.509 PKI 的签名和验证机制来确定一个 MIDlet suites 的可信任性。

1.4 本章小结

本章是全书的理论基础，对 Java 2 平台的体系结构以及它的 3 个版本都做了简要概述，首先解释了 Java ME 在该体系结构中的位置，接下来介绍了 Java ME 的组成结构：虚拟机、配置和简表。

读者需要了解的是关于 Java ME 的应用范围以及它和 CLDC/MIDP 之间的关系，需要重点掌握的是关于 MIDP 的内容，因为它是针对目前手机平台的。第 2 章将讲解如何编写简单的 MIDP 程序和最基本的开发流程。

2 搭建开发平台——Eclipse

本章主要介绍 Java ME 开发环境的搭建。读者需要掌握以下知识点：
- JDK 1.6 安装。
- Eclipse 3.2 安装。
- EclipseMe 1.7.8 安装和配置。
- WTK 2.5 安装和配置。
- MIDLet 项目创建。
- MIDLet 类的创建和运行
- MIDLet 项目打包。

2.1 初识 Eclipse、EclipseME、WTK

2.1.1 Eclipse

Eclipse 是一个开放源代码的、基于 Java 的可扩展开发平台。Eclipse 本身只是一个框架和一组相应的服务，并不能够开发什么程序。在 Eclipse 中几乎每样东西都是插件，实际上正是运行在 Eclipse 平台上的各种插件提供给我们开发程序的各种功能。同时各个领域的开发人员通过开发插件，可以构建与 Eclipse 环境无缝集成的工具。Eclipse 的发行版本都已经带有最基本的插件，方便了开发人员。

你可以在 http://www.eclipse.org/downloads/index.php 下载到 Eclipse 的解压安装文件、语言

包以及许多实用工具插件。本书编写时最新版本是 Eclipse SDK 3.7。不过，在这里我们提醒大家，Eclipse 并不是版本越新越好，新版本往往有一些难以解释的 bug，而且一些插件提供商可能还没有来得及提供与之配套的版本。本文将采用 Eclipse SDK 3.2 为大家演示。

2.1.2　EclipseME

既然 Eclipse 在 Java 开发中如此重要，那么我们能否使用 Eclipse 开发手机应用程序呢？当然，这个答案就是 EclipseME。

EclipseME 作为 Eclipse 的一个插件，致力于帮助开发者开发 Java ME 应用程序。EclipseME 并不为开发者提供无线设备模拟器，而将各手机厂商的实用模拟器紧密连接到 Eclipse 开发环境中，为开发者提供一种无缝统一的集成开发环境。

你可以在 http://www.eclipseme.org/ 上得到免费下载的 EclipseME，本书编写时的最新版本是 0.9，同样出于稳定的考虑，我们在这里选用 eclipseme.feature_0.7.7_site.zip 来为大家演示。

2.1.3　其他工具和环境

除了 Eclipse 与 EclipseME 之外，还需要 Java 运行环境和一些手机模拟器来完成整个搭建工作。以下是本节所需的工具列表（按安装顺序列出）：

JDK 1.6.0　　　　　　http://java.sun.com/j2se/1.4.2/download.html
EclipseME 1.7.7　　　http://www.eclipseme.org/
Sun WTK V2.5　　　　http://java sun.com

说明：最后再重复一次，开发工具不一定是版本越新越好，而是要选择一个稳定搭配的组合套件，作者做 Java ME 开发这么多年，一直采用的就是 Eclipse 3.2 + JDK 1.6 + WTK 2.5 + Eclipse ME 1.7.7，这个组合并不一定是最好，但从作者使用多年的经验来看，这个组合很稳定，这也是本书所用的工具和相关版本。

2.2　搭建 Eclipse 移动开发环境

2.2.1　安装 JDK 1.6

JDK 1.6 的下载地址：

http://www.oracle.com/technetwork/java/javase/downloads/jdk6-jsp-136632.html。
下载完后，双击下载的 exe 文件，出现如图 2-1 所示安装界面。

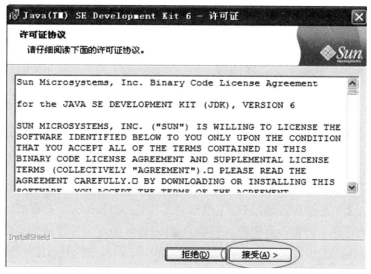

图 2-1　JDK1.6 安装界面

点击"接受",开始安装 JDK,接下来就一直点击"下一步",直到最后结束安装。整个安装过程很简单,但是请最好不要修改 JDK 默认的安装目录,特别是对于新手来说,因为有些开发工具需要的 JDK 环境都是采用的默认安装目录,如果没有找到则需要用户自己去配置相关的环境变量,这对于新手来说比较麻烦,而 JDK 采用默认的安装目录则避免了这个麻烦。

JDK 1.6 安装完毕后出现如图 2-2 所示完成安装的界面。

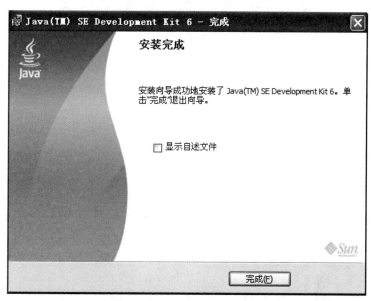

图 2-2　JDK 1.6 安装完毕界面

采用默认的安装路径，则 JDK 的安装目录如图 2-3 所示。

图 2-3　JDK 1.6 默认安装路径

2.2.2　安装 Eclipse

其实，Eclipse 3.2 无需安装，从官网下载后的文件是一个压缩包，直接把这个压缩包解压缩就完成了安装。为了运行方便，请把解压缩后的 `eclipse.exe` 文件发送到桌面快捷方式，运行 `eclipse.exe` 可执行文件后，我们就可以打开 Eclipse 了，双击打开后出现图 2-4 所示界面。

图 2-4　Eclipse 3.2 启动界面

第一次运行 Eclipse 3.2，会出现一个提示对话框，要求你选择项目保存的路径，如图 2-5 所示。

图 2-5　Eclipse 启动选择项目保存路径界面

点击"OK"按钮后，进入 Eclipse 的欢迎页面，关闭这个页面就可以进行项目开发了。如图 2-6 所示就是 Eclipse 的工作界面。

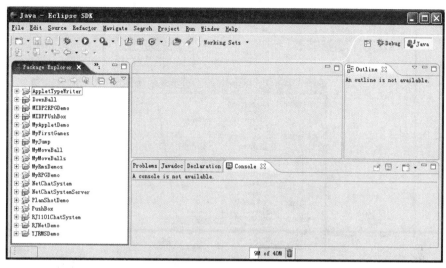

图 2-6　Eclipse 工作界面

2.2.3　安装 EclipseME 插件

在 Eclipse 中选择"Help / Software Updates / find and install..."，在弹出对话框中的选择如图 2-7 所示。

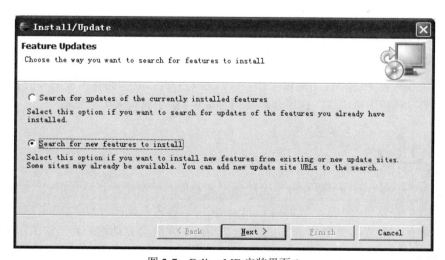

图 2-7　EclipseME 安装界面-1

点击"Next"按钮后，需要选择已经下载好的 EclipseME 插件，选择如图 2-8 所示。

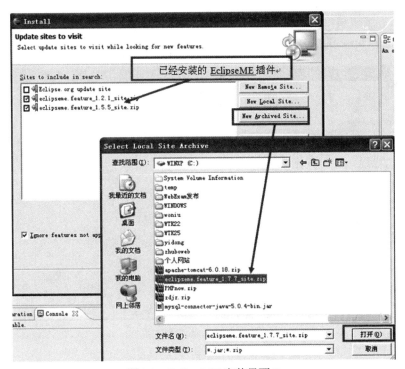

图 2-8　EclipseME 安装界面-2

点击图 2-8 的"打开"按钮后，弹出 EclipseMe 安装界面，后面的安装比较简单，一直点击"Next"和"OK"按钮就能完成，这里不再详细说明。

2.3　加载厂商模拟器

EclipseME 为我们提供了一个集成开发环境，但仅仅靠这些是不够的，我们还需要集成一种或多种手机模拟器来进行程序测试工作。目前，各大手机厂商都拥有多种型号的手机模拟器，Sun 也提供了一种通用模拟器。这里我们采用 Sun WTK 工具包来为大家演示。

加载 Sun WTK v2.5

WTK（Wireless Toolkit）是 Sun 为无线开发者提供的一个无线开发实用包，它拥有多个手机模拟器。我们在这里将 WTK 绑定到 EclipseME，这将大大提高开发者的工作效率。

当然，我们得先安装 WTK。安装过程也很简单，系统会自动检测到当前 JDK 所在路径，并引用该 JDK。

下面将 WTK 绑定到 EclipseME。

找到路径"window/Preference/Java ME/Device Management"，如图 2-9 所示，我们可以添

加当前系统已有的模拟器。在点击"Browse"按钮之后，我们选定 WTK 的安装目录，如图 2-10 所示。

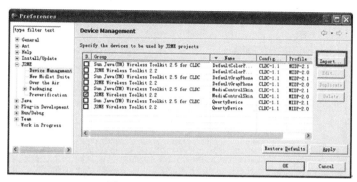

图 2-9　WTK 2.5 安装界面-1

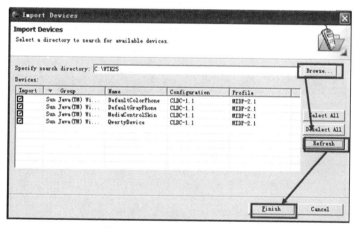

图 2-10　WTK 2.5 安装界面-2

到此，WTK 2.5 安装完毕，现在，我们终于可以用 Eclipse 开发第一个 Java ME 程序了！

2.4　Java ME 项目开发

2.4.1　创建工程

在完成了环境搭建后，我们就可以在 Eclipse 中用我们熟悉的方式开发无线应用程序了。下面来完成一个经典 Hello World 程序。这里，我们选择使用 Sun WTK 2.5 作为模拟器。

选择 Eclipse 界面上菜单栏的"File / New / Other / Java ME Midlet Suite"，如图 2-11 至图 2-13 所示。

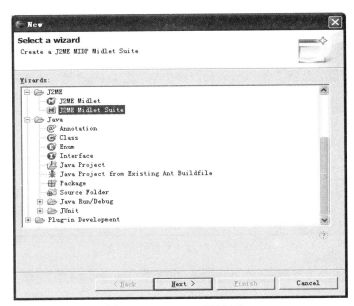

图 2-11 Java ME 项目创建-1

图 2-12 Java ME 项目创建-2

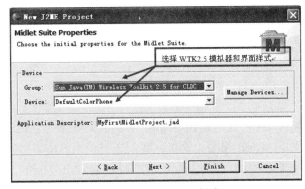

图 2-13 Java ME 项目创建-3

2.4.2 创建 Midlet 类

在完成项目创建后，我们需要创建和运行一个 Midlet 类来查看模拟器执行结果。选择如图 2-14 所示刚创建好的项目，右键单击项目名，选择 "New / Other / Java ME Midlet"，如图 2-15 至图 2-17 所示创建 Midlet 类——MainMidlet。

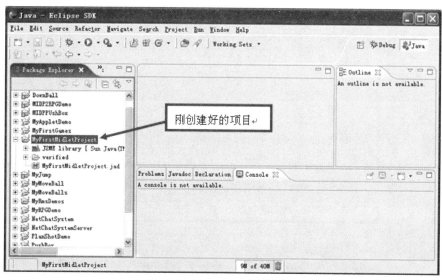

图 2-14　Java ME 项目创建-4

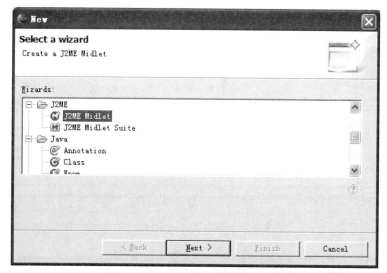

图 2-15　Java ME Midlet 类的创建-1

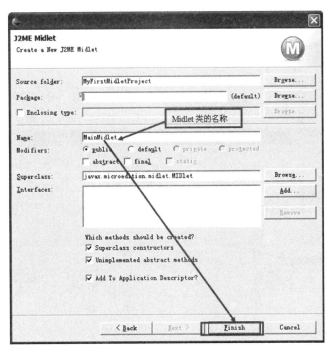

图 2-16　Java ME Midlet 类的创建-2

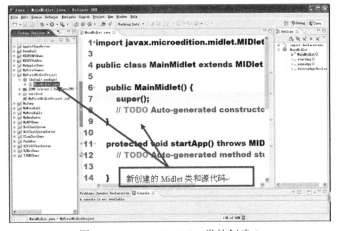

图 2-17　Java ME Midlet 类的创建-3

图 2-17 中创建的 Midlet 类的详细源代码如下（源代码的详细介绍我们留到后面章节再阐述）：

import javax.microedition.lcdui.Display;
import javax.microedition.lcdui.Form;
import javax.microedition.midlet.MIDlet;
import javax.microedition.midlet.MIDletStateChangeException;
public class MainMidlet extends MIDlet {

```java
        Display display = null;//显示对象
        public MainMidlet() {
            super();
            //TODO Auto-generated constructor stub
            display = Display.getDisplay(this);//Display 对象初始化
        }
        protected void startApp() throws MIDletStateChangeException {
            //TODO Auto-generated method stub
            Form form = new Form("MyFirstMIDlet");//Form 类为 Java ME 高级 UI 里面的容器组件
            form.append("我的第一个 Midlet 项目");//在屏幕上输出一行文字
            display.setCurrent(form);//把 Form 作为显示对象显示出来
        }
        protected void pauseApp() {
            //TODO Auto-generated method stub
        }
        protected void destroyApp(boolean arg0) throws MIDletStateChangeException {
            //TODO Auto-generated method stub
        }
    }
```

2.4.3 执行 Midlet

Midlet 类编写完成后，我们就可以执行这个类来查看执行效果。右键单击项目文件夹下刚创建的 MainMidlet.java 类，选择"Run as / Emulated Java ME Midlet"，模拟器运行结果如图 2-18 所示。

图 2-18　Java ME WTK 模拟器运行效果图

2.4.4 打包与混淆

打包，就是为套件生成 jar 文件，用来发布项目。右键单击目标项目，可以在 J2ME 选项中选择 Create Package，生成 jar 包，如图 2-19 所示。

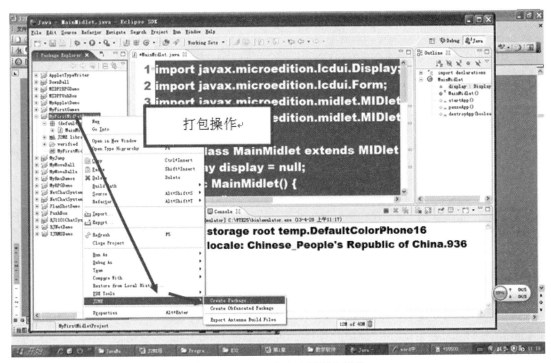

图 2-19　Eclipse 打包和混淆包操作

打混淆包，就是为了保护版权，增加别人反编译阅读源代码的难度；同时可以减少 jar 包的体积。

2.5　本章小结

本章主要介绍了 Java ME 开发工具的安装、相关的环境配置和项目的创建等内容，本书以后所有项目都采用这些开发工具，项目创建也是按照本章介绍的方法和操作顺序，请读者按照本章的操作把自己 PC 上的 Java ME 项目开发工具安装和调试好。

3 MIDP 高级 UI 的使用

本章主要介绍 Java ME 高级 UI 的相关知识和应用。读者需要掌握以下知识点：
- Java ME 的 LCDUI 体系结构。
- Screen 类的结构。
- 各类高级 UI 组件的使用。

3.1 概述

本节要介绍整个 LCDUI 包的结构，让读者对整个 UI 的学习有个大致的了解。图 3-1 为我们展示了整个 LCDUI 包的体系结构。

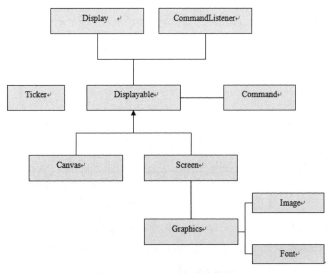

图 3-1　LCDUI 体系结构图

Screen 类属于高级图形用户界面组件，也是我们这一章要着重介绍的内容，Canvas 类是低级图形用户界面组件，在同一时刻，只能有唯一一个 Screen 或者 Canvas 类的子类显示在屏幕上。我们可以调用 Display 类的 setCurrent()方法来将前一个画面替换掉，但必须自行将前一个画面的状态保存起来，并自己控制整个程序画面的切换。

同时可以运用 javax.microedition.lcdui.Command 类提供给我们的菜单项目的功能，分别是：Command.BACKCommand、Command.CANCEL、Command.EXIT、Command.HELP、Command.ITEM、Command.OK、Command.SCREEN 和 Command.STOP。在 Displayable 对象中定义了 addCommand()和 removeCommand()两个方法，这就意味着我们可以在高级 UI 和低级 UI 中同时使用 Command 类，同时可以通过注册 Command 事件来达到事件处理的目的，即 Command 必须与 CommandListener 接口配合使用才能反映用户的动作，具体的使用方法在具体的示例中会给出，读者可以参阅 API 的说明文档获得进一步的认识。

还有在 Displayable 类的子类中都加入了 Ticker，我们可以用 setTicker()来设定画面上的 Ticker，或者用 getTicker()方法来取得画面所含的 Ticker 对象。下面给出 Screen 类的主要结构图，如图 3-2 所示。

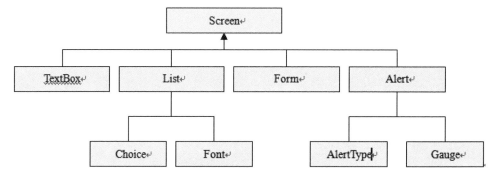

图 3-2　Screen 类结构图

3.2　列表 List

根据 3.1 节的概述我们已经大概了解了 LCDUI 包，现在开始介绍 Screen 类里面的几个重要的类，本节介绍的是 Screen 的一个子类 List，它一共有三种具体的类型：Implicit（简易式）、Exclusive（单选式）、Multiple（多选式）。

与 List 元素相关的应用程序操作一般可概括为 Item 型命令（在后续章节将会有详细介绍）或者 Screen 型命令，其作用域范围的判断依据是看该操作影响到被选择元素还是整个 List 来判定。List 对象上的操作包括 insert、append 和 delete，用于约束 List 具体类型的类是 ChoiceGroup，List 中的元素可以用 getString、insert、set、append、delete、getImage 等方法来具体操纵，对于项目的选择我们则使用 getSelectedIndex()、setSelectedIndex()、 getSelectedFlags()、setSelectedFlags()和 isSelected()来处理，下面来详细介绍一下第一段提到的三个 List 类型。

3.2.1　Exclusive（单选式）

　　和所有的 List 一样，我们可以在构造函数中指定它的标题和类型（构造函数类型 1），也可以使用另一种构造函数类型，即直接传入一个 String 数组和一个 Image 数组，这种构造函数可以直接对 List 内容进行初始化（构造函数类型 2）。在我们进行的大多数开发中，类型 1 的使用是比较常见的，读者可以通过阅读 API 说明文档对其进行深入的掌握。

　　在类型 1 当中，我们需要对其增加内容的时候，就要用到前面提到的 append()方法了，该构造函数的第一个参数是屏幕上的文字，第二个则是代表选项的图标。当不需要图标的时候，和大多数的处理方法相同，只需传入 NULL 这个参数就行了。任何时候我们可以用 insert()方法来插入项目，用 set()方法来重新设置一个项目；当不需要一个项目的时候，可以用 delete()方法来删除特定的选项，我们只需往该方法内传入索引值即可，需要注意的是索引值是从 0 开始，deleteAll()方法则是一次性删除所有的指定 List 的内容。

　　在命令处理函数 commandAction()中，可以用上面提到的几种方法来对用户选择的操作进行侦测，同时定义好对应的处理函数，来达到对应的处理效果。Exclusive 列表如图 3-3 所示。

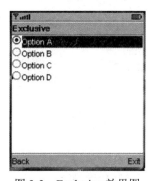

图 3-3　Exclusive 效果图

3.2.2　Implicit（隐含式）

　　implicit（隐含式）其实和上面的单选式没什么大的区别，唯一不同的地方在于命令的处理机制上有一些细微的区别：Choice.Implicit 类型的 List 会在用户选择之后立刻引发事件，并将 List.SelectCommand 作为第一个参数传入。

　　如果不希望该类型的 List 在按下后发出 List.selectCommand 命令作为 commandAction()的第一个参数传入，我们可以用 setSelectCommand(null)将它关掉。需要注意的是，这样做的后果是使 CommandAction()接收到的第一个参数为 null。Implicit 列表如图 3-4 所示。

图 3-4　Implicit 效果图

3.2.3　Multiple（多选式）

Multiple（多选式）类型的 List，顾名思义，可以进行多重选择，其他的地方和上面两种类型大同小异，即可以进行多项的 List 选择如图 3-5 所示。

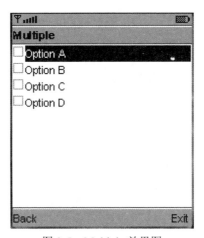

图 3-5　Multiple 效果图

下面我们以 WTK 2.1 自带的 DEMO 为例，通过一段代码来加深巩固这一小节的内容（缺少 API import，请自行添加）：

```
public class ListDemo extends MIDlet implements CommandListener
{
    //这里注意如何使用
    //CommandListener 这个接口
    private final static Command CMD_EXIT = new Command("Exit", Command.EXIT, 1);
    private final static Command CMD_BACK = new Command("Back", Command.BACK, 1);
```

```java
private Display display;
private List mainList;
private List exclusiveList;
private List implicitList;
private List multipleList;
private boolean firstTime;
public ListDemo() {
    display = Display.getDisplay(this);
    String[] stringArray = { "Option A","Option B", "Option C","Option D"};
    //待传入进行初始化的 String 数组，即 Choice 选项的文字部分。
    Image[] imageArray = null;
    //我们这里只是为 Image[]数组进行初始化。
    exclusiveList = new List("Exclusive", Choice.EXCLUSIVE, stringArray, imageArray);
    exclusiveList.addCommand(CMD_BACK); exclusiveList.addCommand(CMD_EXIT); exclusiveList.setCommandListener(this);
    //ExclusiveList 的声明
    implicitList = new List("Implicit", Choice.IMPLICIT, stringArray, imageArray);
    implicitList.addCommand(CMD_BACK); implicitList.addCommand(CMD_EXIT);
    implicitList.setCommandListener(this);
    //ImplicitList 的声明
    multipleList = new List("Multiple", Choice.MULTIPLE, stringArray, imageArray);
    multipleList.addCommand(CMD_BACK); multipleList.addCommand(CMD_EXIT);
    multipleList.setCommandListener(this);
    //MutipleList 的声明
    firstTime = true;
}
protected void startApp() {
    if(firstTime) {
        Image[] imageArray = null;
        try{
            Image icon = Image.createImage("/midp/uidemo/Icon.png");
            //注意！这里的路径是相对路径，请大家千万注意这里的细节问题
            imageArray = new Image[] {icon, icon,icon};
        } catch (java.io.IOException err) {
            //ignore the image loading failure the application can recover.
        }
        String[] stringArray = { "Exclusive", "Implicit","Multiple"};
        mainList = new List("Choose type", Choice.IMPLICIT, stringArray, imageArray);
        mainList.addCommand(CMD_EXIT);
```

```java
                    mainList.setCommandListener(this); display.setCurrent(mainList); firstTime = false;
            }
        }
        protected void destroyApp(boolean unconditional) {

        }
        protected void pauseApp() {

        }
        public void commandAction(Command c, Displayable d) {
            //注意这里是如何实现 CommandListener 这个接口的！
            if (d.equals(mainList)) {
                if (c == List.SELECT_COMMAND) {
                    switch (((List)d).getSelectedIndex()) {
                        case 0:
                            display.setCurrent(exclusiveList);
                            break;
                        case 1:
                            display.setCurrent(implicitList);
                            break;
                        case 2:
                            display.setCurrent(multipleList);
                            break;
                    }
                }else {
                        display.setCurrent(mainList);
                }
            if (c == CMD_EXIT) {
                destroyApp(false);
                notifyDestroyed();
            }
        }
    }
}
```

3.3　TextBox

当我们要在移动设备上输入数据时，TextBox 就派上用场了，如图 3-6 所示。我们使用的 TextBox 的构造函数共有四个参数，第一个是 Title，即标题，第二个是 TextBox 的初始内容，第三个是允许输入字符的最大长度，第四个是限制类型。关于限制类型我们一般按照限制存储内容和限制系统类型分为两种，这两种又各有 6 个具体的类型，大家可以参阅 API 说明文档获得具体类型的运用。在这里我想要提醒读者注意的一点是：一个 TextBox 必须附加一个命令，否则，用户将不能激发任何行为，而陷入这个 TextBox 中。

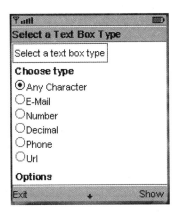

图 3-6 TextBox 效果图

下面给出一个常见的 TextBox 的例子，让大家进一步了解 TextBox：

```
import javax.microedition.lcdui.*;
import javax.microedition.midlet.MIDlet;
public class TextBoxDemo extends MIDlet
implements CommandListener {
    private Display display;
    private ChoiceGroup types; private ChoiceGroup options; private Form mainForm;
    private final static Command CMD_EXIT = new Command("Exit", Command.EXIT, 1);
    private final static Command CMD_BACK = new Command("Back", Command.BACK, 1);
    private final static Command CMD_SHOW = new Command("Show", Command.SCREEN,1);
    /** TextBox的labels
    *
    */
    static final String[] textBoxLabels = {
    "Any Character", "E-Mail", "Number", "Decimal", "Phone", "Url"
    };
    /**
    * 这里列出了几种TextBox的Types
    */
    static final int[] textBoxTypes = {
```

```java
        TextField.ANY, TextField.EMAILADDR, TextField.NUMERIC,
        TextField.DECIMAL, TextField.PHONENUMBER, TextField.URL
};

private boolean firstTime;

public TextBoxDemo() {
    display = Display.getDisplay(this);
    firstTime = true;
}

protected void startApp() {
    if(firstTime) {
    mainForm = new Form("Select a Text Box Type");
    mainForm.append("Select a text box type");

    //the string elements will have no images
    Image[] imageArray = null;

    types = new ChoiceGroup("Choose type", Choice.EXCLUSIVE, textBoxLabels, imageArray);
    mainForm.append(types);

    //进一步选择的选项
    String[] optionStrings = { "As Password", "Show Ticker" };
    options = new ChoiceGroup("Options", Choice.MULTIPLE, optionStrings, null);
    mainForm.append(options); mainForm.addCommand(CMD_SHOW);
    mainForm.addCommand(CMD_EXIT); mainForm.setCommandListener(this); firstTime =false;
    }
    display.setCurrent(mainForm);
}

protected void destroyApp(boolean unconditional) {} /* 抛出异常 throwsMIDletStateChangeException*/
protected void pauseApp() {}
public void commandAction(Command c, Displayable d) {
    if (c == CMD_EXIT) { destroyApp(false); notifyDestroyed();
    } else if (c == CMD_SHOW) {

    //these are the images and strings for the choices. Image[] imageArray = null;
```

```
        int index = types.getSelectedIndex(); String title = textBoxLabels[index]; int choiceType =
    textBoxTypes[index]; boolean[] flags = new boolean[2]; options.getSelectedFlags(flags);
    if (flags[0]) {
    choiceType |= TextField.PASSWORD;
    }
        TextBox textBox = new TextBox(title, "", 50, choiceType);
        if (flags[1]) {
    textBox.setTicker(new Ticker("TextBox: " + title));
    } textBox.addCommand(CMD_BACK); textBox.setCommandListener(this); display.setCurrent(textBox);
    } else if (c == CMD_BACK) {
            display.setCurrent(mainForm);
        }
    }
}
```

3.4 Alert

Alert 类比较有意思，它是用来提醒用户关于错误或者其他异常情况的屏幕对象，这个警告只能作为简短的信息记录和提醒，如果需要长一点的，可以使用其他的 Screen 子类，最常见的是 Form。同时顺便提一下跟它相关的一个类 AlertType，需要提醒读者注意的一点是，AlertType 是一个本身无法实例化的工具类（即我们不能像 Form 那样产生具体的对象）。

AlertType 共有 5 个类型：Alarm（警报），Confirmation（确定），Error（错误），INFO（信息提示），Warning（警告）。

Alert 是一个比较特殊的屏幕对象，当我们在 setCurrent()方法中调用它的时候，它会先发出一段警告的声音，然后才会显示在屏幕上，过了一段时间之后，它会自动跳回之前的画面。

需要注意的是我们必须在使用 setCurrent()显示 Alert 之前定义好它可以跳回的画面，否则会发生异常。

在 Alert 中我们可以通过 setTimeout()方法来设定间隔的时间，setType()来调用上面提到的四种类型，setImage()来定义图片，setString()来定义内含文字，同时通过 getType()，getImage()，getString()来取得相应的对象。

可以利用 setTimeout()来定义 Alert 显示的时间，当 Alert 在屏幕上显示了我们指定的时间间隔后，它会跳回我们指定的屏幕对象，或回到前一个屏幕。如果我们调用 setTimeout()时传入 Alert.Forever 作为参数，那么除非用户按下指定按键，否则，屏幕会一直显示这个 Alert。如果在一个定时的 Alert 中只有一个命令，那么超时发生时命令会自动激活，如图 3-7 所示。

图 3-7　Alert 效果图

```java
import javax.microedition.lcdui.*;
import javax.microedition.midlet.MIDlet;
public class AlertDemo
extends MIDlet {
    private final static Command CMD_EXIT = new Command("Exit", Command.EXIT,1);
    private final static Command CMD_SHOW = new Command("Show", Command.SCREEN,1);
    private final static String[] typeStrings = {
        "Alarm", "Confirmation", "Error", "Info", "Warning"};
    private final static String[] timeoutStrings = {
        "2 Seconds", "4 Seconds", "8 Seconds", "Forever"};
    private final static int SECOND = 1000;
    private Display display;

    private boolean firstTime;
    private Form mainForm;
    public AlertDemo() {
        firstTime = true;
        mainForm = new Form("Alert Options");
    }

    protected void startApp() {
    display = Display.getDisplay(this);
    showOption();
    }

/**
 * 制造这个MIDlet的基本显示
 * 在这个Form里面我们可以选择Alert的各种类型和特性
```

```java
*/
private void showOption() {
if(firstTime) {
//choice-group for the type of the alert:
//"Alarm", "Confirmation", "Error", "Info" or  "Warning"
ChoiceGroup types = new ChoiceGroup("Type", ChoiceGroup.POPUP, typeStrings, null);
mainForm.append(types);

//choice-group for the timeout of the alert:
//"2 Seconds", "4 Seconds", "8 Seconds" or "Forever"
ChoiceGroup timeouts = new ChoiceGroup("Timeout", ChoiceGroup.POPUP, timeoutStrings, null);
mainForm.append(timeouts);
//a check-box to add an indicator to the alert
String[] optionStrings = { "Show Indicator" };
ChoiceGroup options = new ChoiceGroup("Options", Choice.MULTIPLE, optionStrings, null);}
mainForm.append(options); mainForm.addCommand(CMD_SHOW); mainForm.addCommand(CMD_EXIT);
mainForm.setCommandListener(new AlerListener(types, timeouts, options));
firstTime = false;
display.setCurrent(mainForm);
}

private class AlertListener implements CommandListener {

    AlertType[] alertTypes = {
    AlertType.ALARM, AlertType.CONFIRMATION, AlertType.ERROR, AlertType.INFO, AlertType.WARNING
    };
    ChoiceGroup typesCG;
    int[] timeouts = { 2 * SECOND, 4 * SECOND, 8 * SECOND, Alert.FOREVER }; ChoiceGroup timeoutsCG;
    ChoiceGroup indicatorCG;

    public AlertListener(ChoiceGroup types, ChoiceGroup timeouts, ChoiceGroup indicator) {
        typesCG = types;
        timeoutsCG = timeouts;
        indicatorCG = indicator;
    }

    public void commandAction(Command c, Displayable d) {
    if (c == CMD_SHOW) {

    int typeIndex = typesCG.getSelectedIndex(); Alert alert = new Alert("Alert");
```

```java
        alert.setType(alertTypes[typeIndex]);

        int timeoutIndex = timeoutsCG.getSelectedIndex(); alert.setTimeout(timeouts[timeoutIndex]);
        alert.setString(typeStrings[typeIndex]+"Alert,Running"+timeoutStrings[timeoutIndex]);

        boolean[] SelectedFlags = new boolean[1];
        indicatorCG.getSelectedFlags(SelectedFlags);

        if (SelectedFlags[0]) {

            Gauge indicator = createIndicator(timeouts[timeoutIndex]);
            alert.setIndicator(indicator);
        }

            display.setCurrent(alert);
            } else if (c == CMD_EXIT) {
            destroyApp(false);
            notifyDestroyed();
        }
    }
}

protected void destroyApp(boolean unconditional) {}

protected void pauseApp() {}

            /**
             * 我们在这里生成Alert的indicator。
             * 如果这里没有timeout,那么这个indicator 将是一个"非交互性的" gauge
             * 用一个后台运行的thread更新。
             */
            private Gauge createIndicator(int maxValue) {

                if (maxValue == Alert.FOREVER) {

                    return new Gauge(null, false, Gauge.INDEFINITE, Gauge.CONTINUOUS_RUNNING);
                }

                final int max = maxValue / SECOND;
                final Gauge indicator = new Gauge(null, false, max, 0);
```

```
if (maxValue != Gauge.INDEFINITE) {

    new Thread() {
        public void run() {
            int value = 0;

            while (value < max) {
                indicator.setValue(value);
                ++value;

                try { Thread.sleep(1000);
                } catch (InterruptedException ie) {
                //ignore
                }
            }
        }
    }.start();
}

return indicator;
```

3.5 Form 概述

Form 是 Java ME 里面一个比较重要的容器类型，可以说是集中了高级 UI 中的精华，是开发当中常常用到的一个关键类，如图 3-8 所示说明了 Form 及其相关子类的关系。

我们通常是往 Form 里面添加各种 Item 的子类（使用 append()方法），从而达到让画面更加丰富的目的，每一个 Item 的子类在同一时刻只能属于同一个容器，否则会引发异常。

在 Form 画面中，我们通过 Item.LAYOUT_LEFT、Item.LAYOUT_CENTER 和 Item.LAYOUT_RIGHT 来控制各个 Item 在 Form 的位置，通过这几个参数的字面意思可以很容易明白分别是左、中、右。在不设定的情况下，Item 会依照 LAYOUT_DEFAULT 来绘制，如果我们希望自己来设定等效线，可以用 setLayout()方法来控制。

同时，Form 缺省的设定会在空间足够的情况下，尽可能让 Item 出现在同一个逻辑区域中。如果组件在显示时，比我们预期最大尺寸要大（或比预期最小尺寸更小），那么系统会自动忽略之前的设定，转而采用最大尺寸或者最小尺寸，这时系统会自动调用 setPreferredSize()，将预期尺寸设置好。

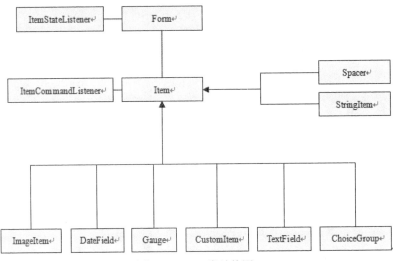

图 3-8 Form 类结构图

3.6　StringItem 及 ImageItem

3.6.1　StringItem

StringItem 的作用，从字面上来看就可以很明白，就是在屏幕上显示一串字。配合不同的外观类型，StringItem 有两个构造函数，最常见的是需要三个参数的，第一个是 Label，第二个是内容，第三个则是外观，外观共分三种：PLAIN、BUTTON、HYPERLINK（只需两个参数的构造函数等同于使用 PLAIN 外观的三个参数的构造函数）。对于外观的提取，我们可以使用 getAppearanceMode()取得，以此类推，需要修改/得到相应的参数只需进行相应的 set/get 操作即可。StringItem 效果如图 3-9 所示。

图 3-9　StringItem 效果图

我们可以把 Item 和其他的高级 UI 部分结合起来，这样也是对我们学习的一种促进：

```java
import javax.microedition.lcdui.*;
import javax.microedition.midlet.MIDlet;

public class StringItemDemo extends MIDlet implements CommandListener,ItemCommandListener {
    private Display display;
    private Form mainForm;
    private final static Command CMD_GO = new Command("Go", Command.ITEM, 1);
    private final static Command CMD_PRESS = new Command("Press", Command.ITEM,1);
    private final static Command CMD_EXIT = new Command("Exit", Command.EXIT, 1);

    protected void startApp() {
        display = Display.getDisplay(this);

        mainForm = new Form("String Item Demo");
        mainForm.append("This is a simple label");

        StringItem item = new StringItem("This is a StringItem label: ", "This is the StringItems text");
        mainForm.append(item);
        item = new StringItem("Short label: ", "text");
        mainForm.append(item);
        item = new StringItem("Hyper-Link ", "hyperlink", Item.HYPERLINK);
        item.setDefaultCommand(CMD_GO);
        item.setItemCommandListener(this);
        mainForm.append(item);
        item = new StringItem("Button ", "Button", Item.BUTTON);
        item.setDefaultCommand(CMD_PRESS);
        item.setItemCommandListener(this);
        mainForm.append(item);
        mainForm.addCommand(CMD_EXIT);
        mainForm.setCommandListener(this);
        display.setCurrent(mainForm);
    }

    public void commandAction(Command c, Item item) {
        if (c == CMD_GO) {
            String text = "Go to the URL...";
            Alert a = new Alert("URL", text, null, AlertType.INFO);
            display.setCurrent(a);
        } else if (c == CMD_PRESS) {
```

```java
                String text = "Do an action...";
                Alert a = new Alert("Action", text, null, AlertType.INFO);
                display.setCurrent(a);
            }
    }

    public void commandAction(Command c, Displayable d) {
        destroyApp(false);
        notifyDestroyed();
    }

    protected void destroyApp(boolean unconditional) {
    }

    protected void pauseApp() {
    }
}
```

3.6.2 ImageItem

下面我们来看 ImageItem。ImageItem 和 StringItem 的区别仅仅在于一个显示图像，一个显示文字，它同样有两个构造函数，其中用到最多的是 5 个参数的构造函数，第一个是该 Item 的 Label，第二个是图片，第三个是等效线，第四个是取代的文字（图片无法显示时），第五个是外观（和 StringItem 相同）。

```java
import javax.microedition.lcdui.*;
import javax.microedition.midlet.*;

public class ImageItemMIDlet extends MIDlet implements ItemCommandListener {
    private Display display;

    public ImageItemMIDlet() {
        display = Display.getDisplay(this);
    }

    public void startApp() {
        Image img = null;
        try {
            img = Image.createImage("/pic.png");
        } catch (Exception e) {
```

```
        }
        Form f = new Form("ImageItem 测试");
        f.append(img);
        ImageItem ii1 = new ImageItem("图片 1", img, Item.LAYOUT_CENTER
                | Item.LAYOUT_NEWLINE_BEFORE, "图片 1 取代文字", Item.BUTTON);
        f.append(ii1);
        ImageItem ii2 = new ImageItem("图片 2", img, Item.LAYOUT_RIGHT
                | Item.LAYOUT_NEWLINE_BEFORE, "图片 2 取代文字", Item.HYPERLINK);
        f.append(ii2);

        display.setCurrent(f);
    }

    public void commandAction(Command c, Item i) {
        System.out.println(c.getLabel());
        System.out.println(i.getLabel());
    }

    public void pauseApp() {
    }

    public void destroyApp(boolean unconditional) {
    }
}
```

3.7 CustomItem

CustomItem 是 Item 中一个比较重要的子类，它最大的优点是提高了 Form 中的可交互性。它和 Canvas 有很大的相似处。我们通过改写 CustomItem 可以实现完全控制在新的子类中条目区域的显示，它可以定义使用的颜色、字体和图形，包括特殊高亮的条目可能有的所有的焦点状态，只有条目的 Label 是由系统控制生成的，但是 Label 总是生成在 CustomItem 的内容区域外。

每个 CustomItem 都负责把它的行为与目标设备上可用的交互模式匹配。一个 CustomItem 调用方法来观察特定设备所支持的交互模式，这个方法会返回一个支持模式的位标记，支持的模式会返回对应位被设置，不支持的不被设置。

CustomItem 一个比较重要的特性即 Form 内部的遍历，即实现可能临时把遍历的责任委派给 Item 本身，这样可以实现特殊的高亮、动画等效果。

由于 CustomItem 在 Form 类里扮演很重要的角色，其内容很庞杂。我们通过三个代码段来教读者如何使用 CustomItem，希望大家通过对代码的深刻认识，提高自己对 CustomItem 的掌握程度。

```java
//第一段代码
import javax.microedition.lcdui.*;
import javax.microedition.midlet.MIDlet;

public class CustomItemDemo extends MIDlet implements CommandListener {
    private final static Command CMD_EXIT = new Command("Exit", Command.EXIT, 1);
    private Display display;

    private boolean firstTime;
    private Form mainForm;

    public CustomItemDemo() {
        firstTime = true;
        mainForm = new Form("Custom Item");
    }

    protected void startApp() {
        if (firstTime) {
            display = Display.getDisplay(this);

            mainForm.append(new TextField("Upper Item", null, 10, 0));
            mainForm.append(new Table("Table", Display.getDisplay(this)));
            mainForm.append(new TextField("Lower Item", null, 10, 0));
            mainForm.addCommand(CMD_EXIT);
            mainForm.setCommandListener(this);
            firstTime = false;
        }
        display.setCurrent(mainForm);
    }

    public void commandAction(Command c, Displayable d) {
        if (c == CMD_EXIT) {
            destroyApp(false);
            notifyDestroyed();
        }
    }

    protected void destroyApp(boolean unconditional) {
```

```java
    }
    protected void pauseApp() {
    }
}
//第二段代码
import javax.microedition.lcdui.*;

public class Table extends CustomItem implements ItemCommandListener {
    private final static Command CMD_EDIT = new Command("Edit", Command.ITEM, 1);
    private Display display;
    private int rows = 5;
    private int cols = 3;
    private int dx = 50;
    private int dy = 20;
    private final static int UPPER = 0;
    private final static int IN = 1;
    private final static int LOWER = 2;
    private int location = UPPER;
    private int currentX = 0;
    private int currentY = 0;
    private String[][] data = new String[rows][cols];

    //Traversal stuff
    //indicating support of horizontal traversal internal to the CustomItem
    boolean horz;

    //indicating support for vertical traversal internal to the CustomItem.
    //boolean vert;

    public Table(String title, Display d) {
        super(title);
        display = d;
        setDefaultCommand(CMD_EDIT);
        setItemCommandListener(this);
        int interactionMode = getInteractionModes();
        horz = ((interactionMode & CustomItem.TRAVERSE_HORIZONTAL) != 0);
        vert = ((interactionMode & CustomItem.TRAVERSE_VERTICAL) != 0);
    }

    protected int getMinContentHeight() {
```

```java
        return (rows * dy) + 1;
}

protected int getMinContentWidth() {

        return (cols * dx) + 1;
}

protected int getPrefContentHeight(int width) {

        return (rows * dy) + 1;
}

protected int getPrefContentWidth(int height) {

        return (cols * dx) + 1;
}

protected void paint(Graphics g, int w, int h) {

        for (int i = 0; i <= rows; i++) {
                g.drawLine(0, i * dy, cols * dx, i * dy);
        }

        for (int i = 0; i <= cols; i++) {
                g.drawLine(i * dx, 0, i * dx, rows * dy);
        }

        int oldColor = g.getColor();
        g.setColor(0x00D0D0D0);
        g.fillRect((currentX * dx) + 1, (currentY * dy) + 1, dx - 1, dy - 1);
        g.setColor(oldColor);

        for (int i = 0; i < rows; i++) {

                for (int j = 0; j < cols; j++) {

                        if (data[i][j] != null) {

                                //store clipping properties int oldClipX = g.getClipX();
                                //int oldClipY = g.getClipY();
                                int oldClipWidth = g.getClipWidth();
```

```java
            int oldClipHeight = g.getClipHeight();
            g.setClip((j * dx) + 1, i * dy, dx - 1, dy - 1);
            g.drawString(data[i][j], (j * dx) + 2, ((i + 1) * dy) - 2,
                    Graphics.BOTTOM | Graphics.LEFT);

            //restore clipping properties

            g.setClip(oldClipX, oldClipY, oldClipWidth, oldClipHeight);
        }
    }
  }
}

protected boolean traverse(int dir, int viewportWidth, int viewportHeight,
        int[] visRect_inout) {

    if (horz && vert) {

        switch (dir) {

        case Canvas.DOWN:

            if (location == UPPER) {
                location = IN;
            } else {

                if (currentY < (rows - 1)) {
                    currentY++;
                    repaint(currentX * dx, (currentY - 1) * dy, dx, dy);
                    repaint(currentX * dx, currentY * dy, dx, dy);
                } else {
                    location = LOWER;

                    return false;
                }
            }

            break;

        case Canvas.UP:

            if (location == LOWER) {
```

```java
                            location = IN;
                        } else {

                            if (currentY > 0) {
                                currentY--;
                                repaint(currentX * dx, (currentY + 1) * dy, dx, dy);
                                repaint(currentX * dx, currentY * dy, dx, dy);
                            } else {
                                location = UPPER;

                                return false;
                            }
                        }

                        break;

                    case Canvas.LEFT:

                        if (currentX > 0) {
                            currentX--;
                            repaint((currentX + 1) * dx, currentY * dy, dx, dy);
                            repaint(currentX * dx, currentY * dy, dx, dy);
                        }

                        break;

                    case Canvas.RIGHT:

                        if (currentX < (cols - 1)) {
                            currentX++;
                            repaint((currentX - 1) * dx, currentY * dy, dx, dy);
                            repaint(currentX * dx, currentY * dy, dx, dy);
                        }
                }
            } else if (horz || vert) {
                switch (dir) {

                case Canvas.UP:
                case Canvas.LEFT:

                    if (location == LOWER) {
                        location = IN;
```

```
            } else {

                if (currentX > 0) {
                    currentX--;
                    repaint((currentX + 1) * dx, currentY * dy, dx, dy);
                    repaint(currentX * dx, currentY * dy, dx, dy);
                } else if (currentY > 0) {
                    currentY--;
                    repaint(currentX * dx, (currentY + 1) * dy, dx, dy);
                    currentX = cols - 1;
                    repaint(currentX * dx, currentY * dy, dx, dy);
                } else {
                    location = UPPER;
                    return false;
                }
            }

            break;

        case Canvas.DOWN:
        case Canvas.RIGHT:
            if (location == UPPER) {
                location = IN;
            } else {

                if (currentX < (cols - 1)) {
                    currentX++;
                    repaint((currentX - 1) * dx, currentY * dy, dx, dy);
                    repaint(currentX * dx, currentY * dy, dx, dy);
                } else if (currentY < (rows - 1)) {
                    currentY++;
                    repaint(currentX * dx, (currentY - 1) * dy, dx, dy);
                    currentX = 0;
                    repaint(currentX * dx, currentY * dy, dx, dy);
                } else {
                    location = LOWER;
                    return false;
                }
            }

        }
    } else {
```

```java
            //In case of no Traversal at all: (horz|vert) == 0
        }

        visRect_inout[0] = currentX;
        visRect_inout[1] = currentY;
        visRect_inout[2] = dx;
        visRect_inout[3] = dy;

        return true;
    }

    public void setText(String text) {
        data[currentY][currentX] = text;
        repaint(currentY * dx, currentX * dy, dx, dy);
    }

    public void commandAction(Command c, Item i) {

        if (c == CMD_EDIT) {

            TextInput textInput = new TextInput(data[currentY][currentX], this,
                    display);
            display.setCurrent(textInput);
        }
    }
}
//第三段代码
import javax.microedition.lcdui.*;
import javax.microedition.midlet.MIDlet;

public class TextInput extends TextBox implements CommandListener {
    private final static Command CMD_OK = new Command("OK", Command.OK,1);
    private final static CommandCMD_CANCEL = new Command("Cancel", Command.CANCEL,1);
    private Table parent;
    private Display display;
    public TextInput(String text, Table parent, Display display) {
        super("Enter Text", text, 50, TextField.ANY);
        this.parent = parent; this.display = display; addCommand(CMD_OK);
        addCommand(CMD_CANCEL);
        setCommandListener(this);
    }
    public void commandAction(Command c, Displayable d) {
```

```
                if (c == CMD_OK) {
                    //update the table's cell and return parent.setText(getString()); display.setCurrentItem(parent);
                } else if (c == CMD_CANCEL) {
                    //return without updating the table's cell display.setCurrentItem(parent);
                }
        }
}
```

3.8 TextField 和 DateField

TextField 和我们前面讲的 TextBox 大同小异，只不过它是作为 Form 的一个子类存在，而 TextBox 则是和 Form 平起平坐，因此我们直接给出代码方便大家的学习。TextField 效果图如图 3-10 所示。

图 3-10　TextField 效果图

```
import javax.microedition.lcdui.*;
import javax.microedition.midlet.*;
public class TextFieldWithItemStateListenerMIDlet extends MIDlet implements ItemStateListener
{
private Display display;
public TextFieldWithItemStateListenerMIDlet()
{
display = Display.getDisplay(this);
}
TextField name ; TextField tel ; TextField summary ; public void startApp()
{
Form f = new Form("TextField测试") ;
name = new TextField("姓名","",8,TextField.ANY) ;
tel = new TextField("电话","",14,TextField.PHONENUMBER) ;
summary = new TextField("总结","",30,TextField.UNEDITABLE) ;
```

f.append(name) ; f.append(tel) ; f.append(summary) ; f.setItemStateListener(this); display.setCurrent(f);
}
public void itemStateChanged(Item item)
{
if(item==name)
{
summary.setString("输入的姓名为： "+name.getString());
}else if(item==tel)
{
public void pauseApp()
{
}
public void destroyApp(boolean unconditional)
{
}
}

DateField 的目的是方便用户输入时间，它的构造函数共有三个参数，一个是 Label，一个是输入模式，一个是 java.util.TimeZone 对象，也可以省去第三个参数，只使用前两个。

3.9 Gauge 和 Spacer，ChoiceGroup

3.9.1 Gauge

Alert 有一套方法可以显示进度，利用 setIndicator()/getIndicator()这组函数，可以显示进度的画面。Gauge 的最大用处就是拿来当进度显示使用如图 3-11 所示。

图 3-11　Gauge 效果图

拿来当进度显示用的 Gauge 对象必须满足如下要求：
● 控制与用户交互的构造函数的第二个参数必须为 false。

- 不能被其他的 Form 或者 Alert 使用。
- 不能加入 Command。
- 不能有 Label。
- 不能自己设定等效线的位置。
- 不能自己设定组件的大小。

大家可以参考如下的代码：

```java
import javax.microedition.lcdui.*;
import javax.microedition.midlet.*;
public class AlertWithIndicatorMIDlet extends MIDlet implements CommandListener {
    private Display display;

    public AlertWithIndicatorMIDlet() {
        display = Display.getDisplay(this);
    }
    Gauge g;
    public void startApp() {
        Alert al = new Alert("处理中");
        al.setType(AlertType.INFO);
        al.setTimeout(Alert.FOREVER);
        al.setString("系统正在处理中");
        g = new Gauge(null, false, 10, 0);
        al.setIndicator(g);
        Command start = new Command("开始", Command.OK, 1);
        Command stop = new Command("停止", Command.STOP, 1);
        al.addCommand(start);
        al.addCommand(stop);
        al.setCommandListener(this);
        display.setCurrent(al);
    }
    public void commandAction(Command c, Displayable s)
    {
        String cmd = c.getLabel();
        if (cmd.equals("开始")) {
            for (int i = 0; i < 11; i++) {
                g.setValue(i);
                try {
                    Thread.sleep(500);
                } catch (Exception e) {
                }
```

```
                }
            } else if (cmd.equals("停止")) {
                notifyDestroyed();
            }
        }

        public void pauseApp() {
        }
        public void destroyApp(boolean unconditional) {
        }
    }
```

3.9.2　Spacer

Spacer 的用处很简单，就是加一处空白，大家可以参考 API 文档进行实际开发。

3.9.3　ChoiceGroup

ChoiceGroup 和 List 大同小异，因为二者都实现了 Choice 接口，所以在很多地方是一样的。但是请注意一点，在这里我们不能使用 Choice.IMPLICIT 类型，只能用 Choice.EXCLUSIVE、Choice.MUTIPLE、Choice.POPUP 三种类型，与 List 的区别即多了第三种弹出式菜单。

3.10　本章小结

本章主要介绍了 Java ME 高级用户界面中常用的一些 UI 组件类以及这些类的的使用方法。本章的内容属于 Java ME 图形界面编程的基础内容，后面章节的实验都以此为基础，所以读者需要认真学习本章内容并能熟练运用。

4
MIDP 低级 UI 的使用

本章主要介绍 Java ME 的相关背景知识。读者需要掌握以下知识点：
- 低级 UI 的使用。
- 低级事件和事件处理。
- 手机上的坐标系体系。
- Image 类的应用。

我们从 javax.microedition.lcdui.Canvas 开始了解低级 UI，要用到低级 UI 必须要继承 Canvas 这个抽象类，而 Canvas 类的核心是 paint() 方法，这个方法就是负责绘制屏幕上的画面，每当屏幕需要重新绘制时，就会产生重绘事件，系统就会自动调用 paint()，并传入一个 Graphics 对象。任何时候我们都可以通过调用 repaint() 方法来产生重绘事件，它有两个方法，一个需要四个参数，分别用来指示起始坐标（X,Y）和长宽，另一个则不需要任何参数，代表重新绘制整个画面。我们可以通过 getWidth() 和 getHeight() 方法获得 Canvas 的当前范围大小。每当 Canvas 范围大小发生变化时，就会自动调用 Canvas 类的 sizeChanged() 方法。

在低级 UI 里，我们可以直接把 Graphics 渲染到屏幕上，也可以在屏幕外合成到一个 Image 中，已渲染的图形具体是合成 Image 还是显示到屏幕上，要看这个 Graphics 具体的来源而定，而渲染到屏幕上的 Graphics 对象将被送到 paint() 方法中来进行调度，这也是显示在屏幕上的唯一的途径。仅在 paint() 方法的执行期间应用程序可以对 Graphics 进行操作，至于要渲染到 Image 中的 Graphics 对象，当需要调用它的时候，可以通过 Image.getGraphics() 方法来取得相应的 Graphics，它将可以被应用程序一直占有，在 paint() 方法运作的任何时候渲染到屏幕上，这也为我们在对不支持 DoubleBuffered 的手机上开发提供了一些思路，即可以通过 Image 来自行设计双缓冲区，避免图像出现所谓的撕裂现象。

4.1 低级 API 与低级事件响应

与高级 UI 相比，低级 UI 就自由很多，任何时候我们可以调用 repaint()产生重绘事件，调用完了 repaint()会立刻返回，调用 paint()回调函数则是由另一个专门的线程来完成。

底层事件大致可分为三类：PressEvents（按键事件），ActionKeys（动作按键），PointerEvents（触控事件）。本节我们将围绕这三类事件来介绍一下底层事件的用法，按键事件的几个核心方法为：

- keyPressed()
- keyReleased()
- keyRepeated()

当按键按下时会触发 keyPressed()，当松开按键时会触发 keyReleased()，当长时间按住按键时会触发 keyRepeated()，但是 RepeatEvents 不是 JTWI 要求强制支持的，所以使用之前要进行测试，看设备是否支持。在 Canvas 里面我们每按下一个按键都会触发 keyPressed()方法，并传入相应位置的整数值。我们在 MIDP 规范中可以很容易发现，KEY_NUM0～KEY_NUM9 十个常数分别代表键盘上的 0～9，还有两个功能键 KEY_STAR，KEY_POUND，如果我们传入的值小于 0，代表我们传入了不合法的 keycode，某些机器还支持连续按键响应，但这并不是 JTWI 规定必须支持的，所以我们在进行实际开发之前一定要用 hasRepeatEvents()方法来进行判定。

动作按键主要是针对游戏来设计的，在 API 中定义了一系列的动作事件：UP、DOWN、LEFT、RIGHT、GAME_A、GAME_B、GAME_C、GAME_D，当按下这些按键时会映射到我们自己为每个按键事件编写的方法，来完成一些动作。不过在 MIDP 2.0 里我们已经有专门的游戏开发包了，所以这里就不重点介绍了。

触控事件主要面向高端设备，并非 JTWI 要求强制支持的，其核心方法为：

- pointerPressed()
- pointerReleased()
- pointerDragged()

分别对应我们通常所用的移动设备手写笔的点击、放开和拖拽三个动作，我们在这三个方法里可以定义相应的事件处理函数。在索爱 P910C 手机上，支持屏幕的触控事件，我们在屏幕上点击，可以触发 pointerPressed()函数，并传入当时位置的坐标；放开后，会触发 pointerReleased()函数，同样也会传入坐标，具体的使用方法和 keyPressed()以及 keyReleased()大同小异。在后面的章节将会有对键盘及触控屏幕事件的详细叙述，同时大家可以参考一下 WTK 的说明文档来获得比较详细的方法和使用规则。

4.2 重绘事件及 Graphics

4.2.1 坐标概念

我们在 MIDP 程序设计中用到的坐标系和平时用到的坐标系不一样，如图 4-1 所示。

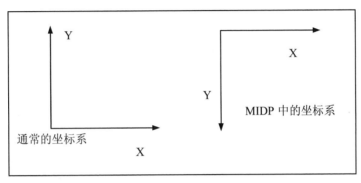

图 4-1　手机坐标系

这是我们在绘制图像时要注意的。

下面来讲一讲 Graphics 这个对象，我们可以把它当作一张白纸，只要调用相应的方法，就可以运用自己的想象力在这张白纸上画出自己想要的图案。

4.2.2 颜色操作

我们在 WTK 的控制台可以看到程序运行以后显示其 Canvas 的 RGB 值和灰度等参数，读者可以运行下面这个程序来获得对 Graphics 个对象的初步了解，如图 4-2 所示。

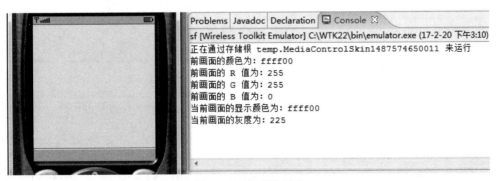

图 4-2　手机颜色操作

下面用一段简单的代码来说明 Graphics 对象的应用：

```java
import javax.microedition.lcdui.Canvas;
import javax.microedition.lcdui.Graphics;

public class test extends Canvas {
    public void paint(Graphics g) {
        g.setColor(255, 255, 0);
        g.fillRect(0, 0, getWidth(), getHeight());
        int c = g.getColor();
        int dc = g.getDisplayColor(g.getColor());
        System.out.println("前画面的颜色为：" + Integer.toHexString(c));
        System.out.println("前画面的 R 值为：" + g.getRedComponent());
        System.out.println("前画面的 G 值为：" + g.getGreenComponent());
        System.out.println("前画面的 B 值为：" + g.getBlueComponent());
        System.out.println("当前画面的显示颜色为：" + Integer.toHexString(dc));
        System.out.println("当前画面的灰度为：" + g.getGrayScale());
    }
}
```

需要大家注意的是 R、G、B 的值只能在 0～255 之间，不可以超出这个范围，另外可以直接用 0x00RRGGBB 格式进行颜色的调配。

4.2.3 绘图操作

Graphics 类提供的大量的绘图操作，这里给出了相关操作的方法列表供读者参考：

方法	功能
void	drawArc(int x, int y, int width, int height, int startAngle, int arcAngle) Draws the outline of a circular or elliptical arc covering the specified rectangle, using the current color and stroke style.
void	drawChar(char character, int x, int y, int anchor) Draws the specified character using the current font and color.
void	drawChars(char[] data, int offset, int length, int x, int y, int anchor) Draws the specified characters using the current font and color.
void	drawImage(Image img, int x, int y, int anchor) Draws the specified image by using the anchor point.
void	drawLine(int x1, int y1, int x2, int y2) Draws a line between the coordinates (x1,y1) and (x2,y2) using the current color and stroke style.
void	drawRect(int x, int y, int width, int height)

	Draws the outline of the specified rectangle using the current color and stroke style.
void	drawRegion(Image src,int x_src, int y_src, int width, int height, int transform, int x_dest, int y_dest, int anchor) Copies a region of the specified source image to a location within the destination, possibly transforming (rotating and reflecting) the image data using the chosen transform function.
void	drawRGB(int[] rgbData, int offset, int scanlength, int x, int y, int width, int height, boolean processAlpha) Renders a series of device-independent RGB+transparency values in a specified region.
void	drawRoundRect(int x, int y, int width, int height, int arcWidth, int arcHeight) Draws the outline of the specified rounded corner rectangle using the current color and stroke style.
void	drawString(String str, int x, int y, int anchor) Draws the specified String using the current font and color.
void	drawSubstring(String str, int offset, int len, int x, int y, int anchor) Draws the specified String using the current font and color.
void	fillArc(int x, int y, int width, int height, int startAngle, int arcAngle) Fills a circular or elliptical arc covering the specified rectangle.
void	fillRect(int x, int y, int width, int height) Fills the specified rectangle with the current color.
void	fillRoundRect(int x, int y, int width, int height, int arcWidth, int arcHeight) Fills the specified rounded corner rectangle with the current color.
void	fillTriangle(int x1, int y1, int x2, int y2, int x3, int y3) Fills the specified triangle will the current color.

主要是一组与绘画和填充有关的方法。请参考 API 查看更多细节。

下面我们就来介绍 Graphics 中线形的概念。如果我们需要绘制一条直线，可以调用 drawLine()方法，需要定义其开始坐标和结束坐标，共四个参数。同时，Graphics 提供两种形式的线条，一个是虚线，即 Graphics.DOTTED，一个是实线，即 Graphics.SOLID，如图 4-3 所示。

图 4-3　简单图形绘制-1

用类似的方法，我们可以实现用 Graphics 的 drawRect() 和 drawRoundRect() 方法来绘制矩形和圆角矩形，如图 4-4 所示，大家仔细观察一下两种矩形的区别。

```java
import javax.microedition.lcdui.*;
import javax.microedition.midlet.*;
public class test3 RectTestCanvas extends Canvas
{
    public void paint(Graphics g)
    {
        clear(g) ;
        g.setColor(255,0,0) ;
        g.drawRect(5,5,100,20);
        g.setColor(0,255,0) ;
        g.fillRect(5,30,100,20);
        //fillRect()和 drawRect()方法的区别在于一个填充一个不填充
        g.setColor(0,0,255);
        g.drawRoundRect(5,55,100,20,20,20);
        g.setColor(255,0,255); g.fillRoundRect(5,80,100,20,20,20);
    }
    public void clear(Graphics g)
    {
        //把屏幕清成白色
        g.setColor(255,255,255);
        g.fillRect(0,0,getWidth(),getHeight());
    }
}
```

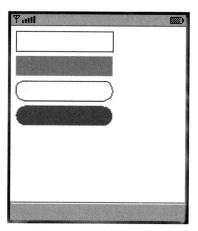

图 4-4　简单图形绘制-2

4.3 Canvas 与屏幕事件处理

Canvas 本身有两种状态，一种是普通默认情况下的，一种是全屏状态下的，可以用 setFullScreenMode()方法来对其设定，两者之间的区别在于当使用全屏幕状态的时候，Title、Ticker 以及 Command 都无法在屏幕上显示。当我们调用 setFullScreenMode()的时候，不管是什么模式，都会调用 seizeChanged()方法，并传入屏幕的高度和宽度作为参数。

对于某些突发事件，比如来电话等，屏幕会被系统画面所覆盖，这时就会调用 hideNotify()方法；当恢复原状时，就会调用原本的画面，那么系统就会同时调用 showNotify()方法。在实际操作过程当中，应该覆写这两个方法，以便在可见性变化时，使程序做出相应的反应。Canvas 会在它被显示的时候自动调用 paint()方法，所以我们不必去调用 repaint()方法。Canvas 事件全屏测试图如图 4-5 至图 4-7 所示。

图 4-5　Canvas 事件全屏测试图 1

图 4-6　Canvas 事件全屏测试图 2

图 4-7　Canvas 事件全屏测试图 3

下面给出一段代码,让大家体会一下如何在实际开发过程中妥善处理屏幕事件:

```java
import javax.microedition.lcdui.Canvas;
import javax.microedition.lcdui.Command;
import javax.microedition.lcdui.CommandListener;
import javax.microedition.lcdui.Displayable;
import javax.microedition.lcdui.Graphics;
import javax.microedition.lcdui.Ticker;

public class test4 extends Canvas implements CommandListener {
    public test4() {
        setTitle("全屏幕测试");
        setTicker(new Ticker("Ticker "));
        addCommand(new Command("全屏幕", Command.SCREEN, 1));
        addCommand(new Command("正常", Command.SCREEN, 1));
        setCommandListener(this);
    }

    public void paint(Graphics g) {
        g.setColor(125, 125, 125);//灰色
        g.fillRect(0, 0, getWidth(), getHeight());
        g.setColor(0, 0, 0);//黑色
        g.drawLine(10, 10, 150, 10);
    }

    public void commandAction(Command c, Displayable s) {
        String cmd = c.getLabel();
        if (cmd.equals("全屏幕")) {
            setFullScreenMode(true);
        } else if (cmd.equals("正常")) {
            setFullScreenMode(false);
        }
    }

    protected void sizeChanged(int w, int h) {
        System.out.println("改变后的宽度:" + w);
        System.out.println("改变后的高度:" + h);
    }

    protected void hideNotify() {
        System.out.println("屏幕被系统遮蔽");//会在 WTK 控制台中显示,
```

```
        //读者需要注意
    }
    protected void showNotify() {
        System.out.println("屏幕显示在屏幕上");
    }
}
```

从图 4-5 和图 4-7 可以看出全屏幕和普通模式的区别,全屏幕的 Canvas 的显示区域覆盖了原来显示标题和 Ticker 的地方。

4.4 键盘及触控屏幕事件的处理

如果需要在 Canvas 里处理按键事件,必须覆写 Canvas 的 keyPressed()、keyReleased()和 keyRepeated()三个方法,其中 keyRepeated()方法 JTWI 并未做硬性规定,所以在开发的时候一定要用 Canvas.hasRepeatedEvents()方法来进行实际的侦测,当按下按键时会触发 keyPressed() 方法,松开时会触发 keyReleased()方法,长时间按住的话则会触发 keyRepeated()方法。JTWI 硬性规定 MIDP2.0 的目标设备必须硬性支持 ITU-T 的电话键盘,即必须使数字 0 到 9、"*"、"#"能在 Canvas 中得到定义,当然也可以扩充其他按键,但是这样对程序的移植就会有影响。下面我们通过代码来看看如何在实际开发中运用上面提到的三个方法。屏幕按键按下前如图 4-8 所示,按下后如图 4-9 所示。

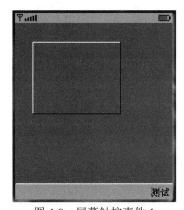
图 4-8　屏幕触控事件-1

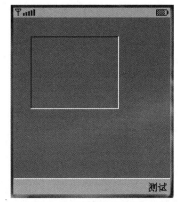
图 4-9　屏幕触控事件-2

代码:

import javax.microedition.lcdui.Canvas;
import javax.microedition.lcdui.Command;
import javax.microedition.lcdui.CommandListener;
import javax.microedition.lcdui.Displayable;

```java
import javax.microedition.lcdui.Graphics;

public class test5 extends Canvas implements CommandListener {
    public test5() {
        addCommand(new Command("测试", Command.SCREEN, 1));
        setCommandListener(this);
    }

    boolean pressed = false;

    public void paint(Graphics g) {
        g.setColor(125, 125, 125);
        g.fillRect(0, 0, getWidth(), getHeight());
        if (pressed) {
            g.setColor(0, 0, 0);
            g.drawLine(20, 20, 120, 20);
            g.drawLine(20, 20, 20, 100);
            g.setColor(255, 255, 255);
            g.drawLine(120, 20, 120, 100);
            g.drawLine(20, 100, 120, 100);
        } else {
            g.setColor(255, 255, 255);
            g.drawLine(20, 20, 120, 20);
            g.drawLine(20, 20, 20, 100);
            g.setColor(0, 0, 0);
            g.drawLine(120, 20, 120, 100);
            g.drawLine(20, 100, 120, 100);
        }
    }

    public void commandAction(Command c, Displayable s) {
        System.out.println("Command Action");
    }

    protected void keyPressed(int keycode) {
        System.out.println("Key Pressed");
        pressed = true;
        repaint();
    }
```

```
    protected void keyReleased(int keycode) {
        System.out.println("Key Released");
        pressed = false;
        repaint();
    }
}
```

对于触控事件（pointer events），应用程序可以通过覆写 pointerPressed()、pointerReleased() 和 pointerDragged 方法实现，（分别对应于手写笔的按下，松开，拖拽三个动作）其处理过程和按键处理几乎一致，所以这里不赘述。

4.5 Graphics 相关类

本节通过设计一个稍微复杂一点的动画来体现 Graphics 在实际开发中带来的便利。这里给出几段曾经对我们启发很大的代码，通过围绕这几段代码进行分析，来掌握 Graphics 在实际开发当中的作用。

4.5.1 Image 类

前面谈到了双缓冲区问题，下面先就 Image 这个类来谈一谈。

在介绍 Image 之前，先介绍几个比较基础的概念，无论是图像还是文字在 Graphics 中都是通过锚点（anchor points）来控制它们具体的方位。对于 Image 而言有如下几个锚点常量：LEFT，RIGHT，HCENTER，TOP，VCENTER，BOTTOM，BASELINE。其具体位置对应如图 4-10 所示。

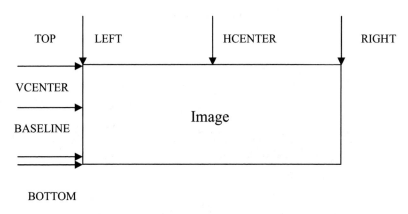

图 4-10 Image 类结构图

Image 分为可变和不可变两种类型。不可变的 Image 是从资源文件、二进制数据、RGB 数值及其他 Image 直接创建的，一旦创建完成，就无法再变化。不可变的 Image 通过 Image.createImage(String name)方法从指定的路径中读取需要创建 Image 所必须的数据，注意参数中的字符串必须以"/"打头，并且包括完整的名称。

可变的 Image 以给定的大小创建，它是可以修改的，可变的 Image 由 Image.createImage(int width,int height)方法创建，需要给定长宽，Image 的其他显示特性和机器的显示屏完全一致。

我们前面提到了撕裂现象，它产生的原因是显示屏在显示图像前都会先参照屏幕虚拟内存，当绘制速度慢到一定程度时，显示在屏幕上的画面会由前一帧画面的一部分和后一帧画面的一部分组成，这样造成了图像"撕裂"。为了解决这个问题，手机厂商可以从硬件上支持 DoubleBuffer，即双缓冲区，这样显示屏绘图的时候就可以使用一个屏幕虚拟内存来绘图，另一个屏幕虚拟内存来进行程序绘图，交叉进行，可以避免画面撕裂的产生。如果手机厂商并未在硬件上支持 DoubleBuffer 该怎么办呢？我们可以从程序上着手，自己设计一个双缓冲区，如图 4-11 所示。

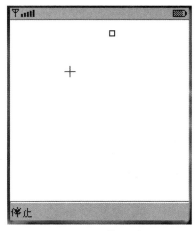

图 4-11　屏幕双缓冲

代码片段如下：

```
import javax.microedition.lcdui.Canvas;
import javax.microedition.lcdui.Command;
import javax.microedition.lcdui.CommandListener;
import javax.microedition.lcdui.Displayable;
import javax.microedition.lcdui.Graphics;
import javax.microedition.lcdui.Image;

public class test6 extends Canvas implements Runnable, CommandListener {
    Command start = new Command("开始", Command.OK, 1);
```

```java
Command stop = new Command("停止", Command.STOP, 1);
private Image offscreen;
public test6() {
    addCommand(start);
    setCommandListener(this);
    if (isDoubleBuffered()) {
        System.out.println("支持双缓冲区");
    } else {
        System.out.println("不支持双缓冲区，启动自制双缓冲区");
        offscreen = Image.createImage(getWidth(), getHeight());
    }
}

public void paint(Graphics g) {
    if (isDoubleBuffered()) {
        System.out.println("On-Screen 绘图");
        clear(g);
        paintAnimation(g, 100, 10, r);
        paintCross(g, x, y, length);
    } else {
        System.out.println("Off-Screen 绘图");
        Graphics offg = offscreen.getGraphics();
        clear(offg);
        paintAnimation(offg, 100, 10, r);
        paintCross(offg, x, y, length);
        g.drawImage(offscreen, 0, 0, 0);
    }
}

public void clear(Graphics g) {
    //把屏幕清成白色
    g.setColor(255, 255, 255);
    g.fillRect(0, 0, getWidth(), getHeight());
}

int r = 0;

public void paintAnimation(Graphics g, int x, int y, int l) {
    g.setColor(0, 0, 0);
```

```java
            g.drawRect(x, y, l, l);
    }
    int x = 50;
    int y = 50;
    int length = 5;
    public void paintCross(Graphics g, int x, int y, int length) {
            g.setColor(255, 0, 0);
            g.drawLine(x - length, y, x + length, y);
            g.drawLine(x, y - length, x, y + length);
    }
    boolean conti = false;
    public void commandAction(Command c, Displayable s) {
            String cmd = c.getLabel();
            if (cmd.equals("停止")) {
                    conti = false;
                    removeCommand(stop);
                    addCommand(start);
            } else if (cmd.equals("开始")) {
                    removeCommand(start);
                    addCommand(stop);
                    conti = true;
                    Thread t = new Thread(this);
                    t.start();
            }
    }
    int rate = 50; //每 1/20 秒画一次
    public void run() {
            long s = 0;
            long e = 0;
            long diff = 0;
            while (conti) {
                    s = System.currentTimeMillis();
                    r++;
                    if (r > 10)
                            r = 0;
                    repaint();
                    serviceRepaints();
                    e = System.currentTimeMillis();
                    diff = e - s;
```

```
                if (diff < rate) {
                    try {
                        Thread.sleep(rate - diff);
                    } catch (Exception exc) {
                    }
                }
            }
        }
        protected void keyPressed(int keycode) {
            switch (getGameAction(keycode)) {
            case Canvas.UP:
                y = y - 2;
                break;
            case Canvas.DOWN:
                y = y + 2;
                break;
            case Canvas.LEFT:
                x = x - 2;
                break;
            case Canvas.RIGHT:
                x = x + 2;
                break;
            }
            repaint();
        }
    }
```

需要提醒各位读者注意的是 Image.createImage()非常浪费内存，最好能够重复使用它。

4.5.2 字体类

Graphics 中还提供了对字体的控制方法，每个 Graphics 都有一个 Font 对象与其关联来进行文字的渲染操作，调用其类方法 setFont(null)，即可使字体恢复到默认状态。对于具体的参数，Font 提供了以下常量来控制 Font 的属性：

- 字体大小：SMALL、MEDIUM、LARGE
- 字体外观：PROPORTIONAL、MONOSPACE、SYSTEM
- 字体风格：PLAIN、BOLD、ITALIC、UNDERLINED

通过 charWidth()、charsWidth()、stringWidth()、substringWidth()可获得字符串、字符、字符集合的宽度，垂直方面则可以通过 getHeight()和 getBaselinePosition()方法获得。

当不对 Font 进行设定时，机器会自动从设备中选择最合适的 Font 属性。

4.6 本章小结

本章主要介绍了 Java ME 的低级用户 UI 组件的使用，在 Java ME 里面低级 UI 主要应用于手机游戏界面的实现，当然部分应用软件为了追求精美的画面也会采用低级 UI 来设计与实现。低级 UI 比高级 UI 具有更自由和更强大的图形实现功能，对于使用 Java ME 实现游戏制作的读者来说低级 UI 是必须掌握的内容。

5

MIDP 的数据存储——RMS

本章主要介绍 MIDP 的数据存储技术。读者需要掌握以下知识点：
- RMS 打开和关闭操作。
- RMS 查询、修改、删除等操作。
- RMS 各类接口的应用。

5.1 初识 RMS（Record Management System）

  记得曾经有人说，数据库程序员是世界上最不愁找不到工作的职业了。虽然此话无从考究，不过也从一个方面说明了无论开发什么类型的应用软件，数据库几乎是一个永恒的话题！在 Java 的体系结构里，现在已经有了 JDBC 技术，还有许多由此衍生的概念，许多耳熟能详的术语，如 EJB、JDO 等等。只是这些都是针对桌面平台或者企业用户的，对于处理能力和存储空间都十分有限的无线设备而言，必须有一种特殊的机制与之适应。MIDP 2.0 规范里不支持全面的树型文件系统，但提供了一种数据持久化机制——记录管理系统（Record Management System，RMS）。

  记录管理系统就是一个小型的数据库管理系统，它以一种简单的、类似表格的形式组织信息，并存储起来形成持久化存储，以供应用程序在重新启动后继续使用。

  RMS 提供了 Records（记录）和 Records Stores（记录仓储）两个概念。

  记录仓储（Records Stores）类似于一般关系数据库系统中的表格（Table），它代表了一组记录的集合。在相同 MIDlet 套件中，每个仓储都拥有自己独一无二的名字，大小不能超过 32 个 Unicode 字符，同一个套件下的 MIDlet 都可以共享这些记录仓储。

  记录是记录仓储的组成元素。记录仓储中含有很多条记录，就如同记录表格是由一行行组成的一样。每条记录代表了一条数据信息。一条记录（Record）由一个整型的 RecordID 与一个代表数据的 byte[]数组两个子元素组成。RecordID 是每条记录的唯一标志符，利用这个标志符可以从记录仓储中找到对应的一条记录。请注意，由于产生记录号 RecordID 使用的是一种简单的单增算法。当一条数据记录被分配的时候，它的记录号也就唯一分配了。并且该条记录被删除后，RecordID 也不会被使用。所以，仓储中相邻的记录并不一定会有连续的 RecordID。

MIDP 套件所使用的 RMS 空间图如图 5-1 所示。

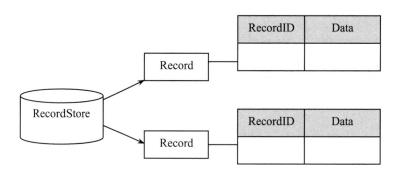

图 5-1 RMS 空间图

每一个 MIDlet 套件都会有属于自己的一个用于 RMS 的私有空间，可以通过 jad 描述文件事先规定 Midlet 套件运行所必需的 RMS 空间大小——在手机内部存储空间中预留的一个空间，供由 jad 指定的 jar 包文件使用。在 MIDP 2.0 以后，只要 MIDlet 开放了属于自己 RMS 空间的相应 RecordStore 的使用权限，那么这个 RecordStore 可以被套件外部的其他 MIDlet 访问。

所有与 RMS 相关的 API 都集中在 javax.microedition.rms 包下，包括了一个主类、四个接口，以及五个可能被抛出的异常。既然 RMS 在结构上就分为记录与记录仓储两个部分，那么对它们的操作当然也是有所区别的，5.2 节"RecordStore 的管理"将首先阐述记录仓储自身的操作。

5.2 RecordStore 的管理

5.2.1 RecordStore 的打开

当你查阅 MIDP API 文档的时候，将略为惊讶的发现，RecordStore 并不能通过 new 来打开或创建一个实例。事实上，RecordStore 提供了一组静态方法 openRecordStore()来取得实例。这里，你有三个选择：

- openRecordStore (String recordStoreName, boolean createIfNecessary)

- openRecordStore (String recordStoreName, boolean createIfNecessary, int authmode, boolean writable)
- openRecordStore (String recordStoreName, String vendorName, String suiteName)

显然，最复杂（也是最灵活的）是第二种打开方法。它的第一个参数是记录仓储的名称，第二个参数表明了当我们请求的仓储不存在时，是否新建一个 RecordStore。第三个参数表明了此仓储的读取权限，最后一个参数则决定了写入权限。

当我们使用第一个打开方法时，则表示我们选择读取权限只限于本地，并且拒绝其他 MIDlet 写数据到这个记录仓储上。即相当于使用第二种打开方法并分别为第三、第四个参数传入了 RecordStore.AUTHMODE_PRIVATE 和 FALSE。

最后，MIDP API 中还提供了一个专门用来读取其他 MIDlet 套件记录仓储的打开方法；它的三个传入参数分别为记录仓储名、发布商名以及 MIDlet 套件名。请注意，如果该记录仓储的读取权限为 AUTHMODE_PRIVATE 的话，此方法将返回安全错误。

下面，我们给出一个使用第一种方法的示例：

```
private RecordStore rs = null;
try {
    //打开一个 RMS，如果打开失败，则创建一个
    rs = RecordStore.openRecordStore("testRMS", true);
} catch (RecordStoreNotFoundException e) {
    e.printStackTrace();
} catch (RecordStoreFullException e) {
    e.printStackTrace();
} catch (RecordStoreException e) {
    e.printStackTrace();
}
```

5.2.2　RecordStore 的关闭

当我们不再使用一个打开的 RecordStore 时，记得要关闭它以节约资源。这时，你就需要使用 RecordStore 的关闭方法 closeRecordStore()。

在记录仓储关闭期间，加载在当前记录仓储上的所有监听器将被消除，记录仓储本身也不能再被调用或遍历，任何试图对 RMS 采取的操作都会抛出 RecordStoreNotOpenException 异常。关闭方法是如此简单，以至于我们有可能忽略这样一个细节：记录仓储在调用 closeRecordStore()方法之后不会立即被关闭，除非你确信关闭方法的调用次数与打开方法 openRecordStore()的调用次数一样多。也就是说，MIDlet 套件需要在 RMS 的打开与关闭之间保持平衡。

下面给出了关闭记录仓储的一个示例：

```
try {
    rs = RecordStore.openRecordStore("testRMS", true);
    //DO YOUR WORK HERE
    rs.closeRecordStore();
} catch (RecordStoreNotOpenException e) {
    e.printStackTrace();
} catch (RecordStoreException e) {
    e.printStackTrace();
}
```

5.2.3　RecordStore 的删除

当我们不再需要一个记录仓储的时候，就需要调用 deleteRecordStore()方法来执行删除。一个 MIDlet 套件只能够删除它自己的记录仓储。在删除之前，我们需要确保当前的仓储处于关闭状态，如果该仓储仍旧处于打开状态（被自身的 MIDlet 或者其他套件调用），那么删除记录将导致 RecordStoreException 异常抛出；如果要删除的仓储记录本身不存在，那么将会引起 RecordStoreNotFoundException 异常抛出。

```
//假定 rs 是已经存在的记录仓储，并已经打开
try {
    rs.closeRecordStore();
    RecordStore.deleteRecordStore("testRMS");
} catch (RecordStoreNotOpenException e) {
    e.printStackTrace();
} catch (RecordStoreNotFoundException e) {
    e.printStackTrace();
} catch (RecordStoreFullException e) {
    e.printStackTrace();
} catch (RecordStoreException e) {
    e.printStackTrace();
}
```

5.2.4　其他相关操作

RecordStore 中除去以上介绍的三种最常用的方法外，还包括了一些很有用的操作，可以取得对记录仓储的信息，包括：
- getLastModified()：返回记录仓储最后更新时间。
- getName()：返回一个已经打开了的记录仓储名称。
- getNumRecords()：返回当前仓储中记录总数。

- getSizeAvailable()：返回当前仓储中可用的字节数。
- getVersion()：返回记录仓储版本号。
- listRecordStores()：获取该 MIDlet 套件中所有的记录仓储列表。

以上所说的都是针对记录仓储的操作，就如同一个关系型数据库系统，RMS 也有包括插入、删除在内的对单条记录的基本操作，这将在下一节中介绍。

5.3 RecordStore 的基本操作

5.3.1 增加记录

我们可以通过方法 addRecord(byte[] data, int offset, int numBytes)添加 byte 数组类型的数据。在 addRecord 中，我们需要提供三个有效参数：byte[]数组、传入 byte[]数组起始位置、传入数据的长度。

当数据添加成功后，addRecord 将返回记录 ID 号（RecordID），RecordID 在一个 RecordStore 中扮演着主键的角色，它由一个简单增长算法产生。例如第一条添加的记录 ID 是 0，第二条是 1，以此类推……。如果试图将数据添加到一个未经打开的记录仓储中，将产生 RecordStoreNotOpenException 异常，任何添加错误都将最终抛出 RecordStoreException 异常，所以，将它作为最后一个 catch 块将是一个很好的选择。

增加操作是一个阻塞操作，直到数据被持久的写到了存储器上，对 addRecord 方法的调用才会返回。同时这个操作也是一个原子操作，这意味着多个线程同时调用 addRecord 不会产生数据写丢失。但这并不能保证读取和写入同时发生时读取的自动同步，所以复杂应用中对应的同步机制是必须的。

5.3.2 修改与删除记录

通过方法 deleteRecord(int recordId)传入目标记录的 ID 以后，可以从记录仓储中删除记录。需要注意的是，正如前面提过的记录 ID 号是不能够被复用的。如果试图从一个尚未打开的记录仓储中删除记录，将会抛出 RecordStoreNotOpenException 异常；如果传入的记录 ID 是无效的，将得到 InvalidRecordIDException 异常，而 RecordStoreException 异常则可以用来捕获一般性的仓储错误。

通过方法 setRecord(int recordId, byte[] newData, int offset, int numBytes)可以修改一个指定 ID 的记录值。修改记录的传入参数包括记录号，其余的与 addRecord 相同。当仓储没有打开或者传入 ID 无效时，你会得到与 deleteRecord 一致的抛出错误。而 RecordStoreException 异常同样可以用来捕获一般性的仓储错误。

5.3.3 自定义数据类型与字节数组的转换技巧

我们在前面已经提到，针对 RecordStore 的操作只提供对 byte 数组的服务，而在日常处理中，大部分时候我们遇到的都将是非 byte 类型。当然，我们可以编写一些方法来完成基本类型（int, char）与 byte 的相互转换。但我们又将面临如何解决另一种更常见的问题：自定义数据类型与 byte 的转换？在这里，向大家介绍一种已经被广泛采用的方法。

要将自定义数据类型与 byte 数组相互转换，需要 ByteArrayOutputStream、DataOutputStream、ByteArrayInputStream、DataInputStream 四个类的协助。在 MIDP 的帮助 API 中有对于它们的详细描述，你可以在 http://java.sun.com 下载或在线查阅。下面简单介绍一下它们的使用。

要写入数据首先要建立一个 ByteArrayOutputStream 的实例 baos，然后将它作为参数传入 DataOutputStream 的构造函数来产生一个实例 dos。dos 有一组方便的 I/O 方法 writeXXX，方便我们将不同的数据写入流。例如 writeInt 用于写入 int 型、writeChar 用于写入字符型、writeUTF 用于写入一个 String 等。更多信息请参考 API 手册。当写入操作完成后，可以利用 baos 的 toByteArray 方法得到一个 byte[]数组，这个数组含有我们刚刚写入的数据，将它传给 addRecord 就可以增加一条记录了。最后记住要关闭打开的流。

要读入数据就要利用剩余的两个类 ByteArrayInputStream 和 DataInputStream。首先利用 getRecord(int)得到刚刚写入的 byte 数组再利用得到的 byte 数组构造一个 ByteArrayInputStream 的实例 bais，然后用 DataInputStream 包装它，得到一个实例 dis。DataInputStream 有一组方便的 I/O 方法用于读入 DataOutputStream 对应方法写入的数据。应该注意的是读入顺序和写入顺序应保持一致。同样的不再使用流时，也应关闭流以节约资源。

以下是一段代码示范，首先写入一组自定义数据，然后再读出：

```
ByteArrayOutputStream baos=new ByteArrayOutputStream();
DataOutputStream dos=new DataOutputStream(baos);
dos.writeBoolean(false);
dos.writeInt(15);
dos.writeUTF("abcde");
byte [] data=baos.toByteArray();//取得 byte 数组
dos.close();
baos.close();
ByteArrayInputStream bais=new ByteArrayInputStream(data);
DataInputStream dis=new DataInputStream(bais);
boolean flag=dis.readBoolean();
int intValue=dis.readInt();
String strValue=dis.readUTF();
dis.close();
bais.close();
```

5.3.4 利用 RMS 实现对象序列化

有了上一节的基础，现在我们可以很容易地利用 RMS 来实现对象的序列化。下面，以一个单词记录本为示例，为大家展示如何序列化对象。单词记录本的基本类是一个 Word，包括英文单词 enWord、解释 cnWord、上次读取时间 dateTime，以及备注信息 detail。

```java
public class Word {
    private String enWord;
    private String cnWord;
    private long dateTime;
    private String detail;
}
```

我们将数据转换作为 Word 类的内置方法，serialize 用于序列化对象数据，返回 byte 数组类型；deserialize 完成的则是相反的工作。对象中成员变量的读取方法 writeXXX 要与变量的类型相吻合，并且完成操作以后，要将流及时关闭！

```java
/**生成序列化的 byte 数组数据
*/
public byte[] serialize() throws IOException{
    //Creates a new data output stream to write data
    //to the specified underlying output stream
    ByteArrayOutputStream baos = new ByteArrayOutputStream();
    DataOutputStream dos = new DataOutputStream(baos);
    dos.writeUTF(this.enWord);
    dos.writeUTF(this.cnWord);
    dos.writeLong(this.dateTime);
    dos.writeUTF(this.detail);
    baos.close();
    dos.close();
    return baos.toByteArray();
}
/**将传入的 byte 类型数据反序列化为已知数据结构
*/
public static Word deserialize(byte[] data) throws IOException{
    ByteArrayInputStream bais = new ByteArrayInputStream(data);
    DataInputStream dis = new DataInputStream(bais);
    Word word = new Word();
    word.enWord = dis.readUTF();
    word.cnWord = dis.readUTF();
    word.dateTime = dis.readLong();
    word.detail = dis.readUTF();
```

```
            bais.close();
            dis.close();
            return word;
    }
```

5.4 RecordStore 的高级操作

记录仓储类似于一个简单的数据库，不仅仅体现在最基本的添加删除操作上，RecordStore 同样为开发者提供了一些高级功能，这主要体现在四个接口的使用：RecordComparator、RecordEnumeration、RecordFilter 与 RecordListener。考虑到 RecordComparator、RecordFilter 都是作用在 RecordEnumeration 上的，故先来介绍这个接口。

5.4.1 RecordEnumeration 遍历接口

当我们需要遍历一个记录仓储中的所有记录时，通常的想法是使用一个 for 循环，利用 RecordID 来实现遍历。但是很遗憾，在 RMS 中，这不是一个好方法。首先，我们往往无从了解之前存入数据的 RecordID，另外，也是最重要的一点，即便 RecordID 还存在，记录本身也可能已经被删除了，即我们不能保证 RecordID 对应记录的有效性。

这样的话，我们就需要一些更为有效的方法来遍历记录仓储。MIDP 规范中提供了一种安全、可靠的遍历方式——RecordEnumeration 接口。

RecordEnumeration 内部没有存放任何记录仓储数据的副本，因此在使用 RecordEnumeration 读取数据的时候，实际上仍旧是抓取 RecordStore 之中的数据。RecordEnumeration 如同是一个可用 RecordID 的集合，它甚至可以按我们指定的方式排列记录。

下面介绍和 RecordEnumeration 相关的几个主要方法：

- enumerateRecords()

通过对 RecordStore 实例对象调用 enumerateRecords 方法可以取得一个 RecordEnumeration 的实例。

在 enumerateRecords 方法中我们可以传入三个参数：filter、comparator 与 keepUpdated。前两个参数分别是过滤器和排序策略，这个后面会讲到。当传入的 filter 不为空时，它将用于决定记录仓储中的哪些记录被使用。当 comparator 不为空时，RecordStore 将按照我们指定的排列顺序返回。第三个参数决定了当遍历器建立起来以后，是否对记录仓储新的改变做出回应。如果传入 true，那么将有一个 RecordListener 被加入到 RecordStore 中，使得记录仓储的内容与 RecordEnumeration 随时保持同步；如果传入 false，则可以使得当前遍历更有效率，但所取得的 RecordID 集合仅仅是调用此方法这个时刻的 RecordStore 快照，此后对 RecordStore 的所有更改都不会反映在这个集合上。请读者根据要求在访问数据完整性和访问速度之间进行取舍。

- numRecords()

numRecords 返回了在当前遍历集合中可用的记录数目。这里所指的可用,不仅仅是说 RecordID 对应的记录存在;当 filter 存在时,也需要符合过滤条件。

- hasNextElement()

hasNextElement 用于判断在 RecordEnumeration 当前指向的下一个位置还有没有剩余记录。

- hasPreviousElement()

hasPreviousElement 用于判断在 RecordEnumeration 当前指向的前一个位置还有没有剩余记录。

- nextRecord()

nextRecord 返回遍历器下一位置的记录副本,由于返回的是副本,所以任何对返回记录的修改都不会影响到记录仓储的实际内容。

- nextRecordId()

nextRecordId 返回当前遍历器下一位置记录的 RecordID,当下一位置没有可用的记录时,继续调用 nextRecordId 将抛出 InvalidRecordIDException 异常。

依旧以单词本为例,添加一个方法 ViewAll(),用来返回当前记录仓储中的所有单词,演示了如何利用 RecordEnumeration 来遍历记录。在示例中,使用到了 RecordComparator 接口的一个实例 WordComparator,后面会讲到。

```
public Word[] viewAll() throws IOException {
    Word[] words = new Word[0];
    RecordEnumeration re = null;
    //rs 是之前创建的 RecordStore 类型实例变量
    if (rs == null)
    return words;
    try {
        re = rs.enumerateRecords(null, new WordComparator(), false);
        //无过滤器、但有一个排序策略
        words = new Word[re.numRecords()];
        int wordRecords = 0;
        while (re.hasNextElement()) {
            byte[] tmp = re.nextRecord();
            words[wordRecords] = Word.deserialize(tmp);
            wordRecords++;
        }
    } catch (RecordStoreNotOpenException e1) {
        e1.printStackTrace();
    } catch (InvalidRecordIDException e1) {
        e1.printStackTrace();
    } catch (RecordStoreException e1) {
```

```
            e1.printStackTrace();
        } finally {
            if (re != null)
                re.destroy();
        }
        return words;
}
```

5.4.2　RecordFilter 过滤接口

过滤接口是用来过滤不满足条件的记录的。使用 RecordFilter 接口必须实现 match(byte[] candidate)方法，当传入 byte 数据符合筛选条件时，返回 true。

下面的示例建立一个静态类 WordFilter，它实现了 RecordFilter 接口。

```
public class WordFilter implements RecordFilter{
    private String enWord;
    private int type;
    public WordFilter(String enword, int type){
        //传入要比较的项，type 指向一个自定义的内部事件标记
        //表现为整型
        this.enWord = enword;
        this.type = type;
    }
    public boolean matches(byte[] word) {
        //matches 方法中传入的参数是 RMS 中的各个候选值（元素）
        try {
            if(type == EventID.SEARCH_EQUAL){
                return Word.matchEN(word, enWord);
            }else{
                return Word.matchEN_StartWith(word, enWord);
            }
        } catch (IOException e) {
            e.printStackTrace();
            return false;
        }
    }
}
```

以上示例中的 EventID.SEARCH_EQUAL 为一个定义好的整型数据。同时，这里涉及到了 Word 类的两个对应方法：

```
public static boolean matchEN_StartWith(byte[] data, String enword) throws IOException{
    ByteArrayInputStream bais = new ByteArrayInputStream(data);
```

```java
        DataInputStream dis = new DataInputStream(bais);
        try{
            return (dis.readUTF().startsWith(enword));
        }catch(IOException e){
            e.printStackTrace();
            return false;
        }
    }

    public static boolean matchCN(byte[] data, String cnword) throws IOException{
        ByteArrayInputStream bais = new ByteArrayInputStream(data);
        DataInputStream dis = new DataInputStream(bais);
        try{
            dis.readUTF();
            return (dis.readUTF().equals(cnword));
        }catch(IOException e){
            e.printStackTrace();
            return false;
        }
    }
```

5.4.3 RecordComparator 比较接口

比较器定义了一个比较接口，用于比较两条记录是否匹配，或者符合一定的逻辑关系。使用比较器必须实现方法 compare(byte[] rec1, byte[] rec2)，当 rec1 在次序上领先于 rec2 时，返回 RecordComparator.PRECEDES；反之则返回 RecordComparator.FOLLOWS；如果两个传入参数相等，RecordComparator 将返回 RecordComparator.EQUIVALENT。如同过滤器一样，我们设计一个静态类——WordComparator，用以实现 RecordComparator 接口。

```java
import java.io.IOException;
import javax.microedition.rms.RecordComparator;
private static class WordComparator implements RecordComparator {
    public int compare(byte[] word_1, byte[] word_2) {
        try {
            Word word1 = Word.deserialize(word_1);
            Word word2 = Word.deserialize(word_2);
            long dateTime1 = word1.getDateTime();
            long dateTime2 = word2.getDateTime();
            if (dateTime1 < dateTime2) {
                return RecordComparator.FOLLOWS;
```

```
            }
            if (dateTime1 > dateTime2) {
                return RecordComparator.PRECEDES;
            }
            return RecordComparator.EQUIVALENT;
        } catch (IOException e) {
            e.printStackTrace();
        }
        return 0;
    }
}
```

正如你看到的为了使程序更加的优化，我们都是利用序列化/反序列化技术取得完整的对象后，再通过对对象自身方法的调用来实现过滤或者排序的核心逻辑。这样做占用了大量的处理器时间。所以在使用这项技术时，请注意优化你的算法并确保使用它们是必要的。

5.4.4 RecordListener 监听器接口

最后来关注一下 RecordListener 接口。RecordListener 是用于监听记录仓储中记录添加、更改或删除等事件的接口。它是作用在 RecordStore 上的，利用 RecordStore 的 addRecordListener 方法来注册一个监听器。使用监听器必须实现三个方法：recordAdded，recordChanged 与 recordDeleted，它们都需要传入两个参数：记录仓储名称 recordStroe 与记录号 recordId。

- recordAdded：当一条新的记录被添加到仓储空间的时候，该方法被触发。
- recordChanged：当一条记录被修改时使用。
- recordDeleted：当一条记录从记录仓储中删除时调用。

需要注意的是，RecordListener 是在对记录仓储的操作动作完成以后才被调用的！特别在 recordDeleted 方法中，由于传入的记录已经删除，所以如果再使用 getRecord()试图取得刚刚被删除记录的话，将会抛出 InvalidRecordIDException 异常。

5.5 本章小结

本章主要介绍了 MIDP 中数据库的基本应用，在 MIDP 中的数据操作主要是通过 RMS 这个简单数据库来实现。RMS 在 Java ME 游戏中一般用来保存游戏进度、玩家积分数据和其他游戏过程中简单类型的数据。

6

GAME API（MIDP2.0）

本章主要介绍与 MIDP 2.0 游戏开发相关的 API。读者需要掌握以下知识点：
- Layer 类。
- Sprite 类的应用。
- TiledLayer 类的应用。
- LayerManager 类的使用。

6.1 游戏 API 简介

MIDP 2.0 相对于 1.0 来说，最大的变化就是新添加了用于支持游戏的 API，它们被放在 javax.microedition.lcdui.game 包中。游戏 API 包提供了一系列针对无线设备的游戏开发类。由于无线设备仅有有限的计算能力，因此许多 API 的目的在于提高 Java 游戏的性能，并且把原来很多需要手动编写的代码如屏幕双缓冲、图像剪裁等都交给 API 间接调用本地代码来实现。各厂家有相当大的自由来优化它们。

游戏 API 使用了 MIDP 的低级图形类接口（Graphics、Image 等）。整个 game 包仅有五个 Class：
- GameCanvas

这个类是 LCDUI 的 Canvas 类的子类，为游戏提供了基本的"屏幕"功能。除了从 Canvas 继承下来的方法外，这个类还提供了游戏专用的功能，如查询当前游戏键状态和同步图像输出，这些功能简化了游戏开发并提高了性能。
- Layer

Layer 类代表游戏中的一个可视化元素，例如 Sprite 或 TiledLayer 是它的子类，这个抽象类搭好了层（Layer）的基本框架并提供了一些基本的属性，如位置、大小、可视与否等。出于优化的考虑，不允许直接产生 Layer 的子类（即不能包外继承）。

- LayerManager

对于有着许多 Layer 的游戏而言，LayerManager 通过实现分层次的自动渲染，从而简化了游戏开发。它允许开发者设置一个可视窗口（View Window）来表示用户在游戏中可见的窗口，LayerManager 自动渲染游戏中的 Layer，从而实现期望的视图效果。

- Sprite

Sprite 又称"精灵"，也是一种 Layer，可以显示一帧或多帧的连续图像。但所有的帧都是相同大小的，并且由一个 Image 对象提供。Sprite 通过循环显示每一帧，可以实现任意顺序的动画；Sprite 类还提供了许多变换（翻转和旋转）模式和碰撞检测方法，能大大简化游戏逻辑的实现。

- TiledLayer

TiledLayer 又称"砖块"，这个类允许开发者在不必使用非常大的 Image 对象的情况下创建一个大的图像内容。TiledLayer 由许多单元格构成，每个单元格能显示由一个单一 Image 对象提供的一组贴图中的某一个贴图。单元格也能被动画贴图填充，动画贴图的内容能非常迅速地变化，这个功能对于动画中要显示非常大的一组单元格时非常有用，例如一个充满水的动态区域。

在游戏中，某些方法如果改变了 Layer、LayerManager、Sprite 和 TiledLayer 对象的状态，通常并不能立刻呈现出视觉变化。因为这些状态仅仅存储在对象里，只有当随后调用我们自己的 paint()方法时才会更新显示。这种模式非常适合游戏程序，因为在一个游戏循环中，一些对象的状态会更新，但在每个循环的最后，整个屏幕才会被重绘。基于轮询也是现在视频游戏的基本结构。

6.2　GameCanvas 的使用

GameCanvas 类提供了基本的游戏用户接口。除了从 Canvas 类继承下来的特性（命令、输入事件等）以外，它还提供了专门针对游戏的功能，比如后备屏幕缓冲和键盘状态查询的能力。每个 GameCanvas 实例都会有一个为之创建的专用的缓冲区。因为每个 GameCanvas 实例都会有一个唯一的缓冲区，可以从 GameCanvas 实例获得其对应的 Graphics 对象，而且，只有对 Graphics 对象操作，才会修改缓冲区的内容。外部资源如其他的 MIDlet 或者系统级的通知都不会导致缓冲区内容被修改，该缓冲区在初始化时被填充为白色。

缓冲区大小被设置为 GameCanvas 的最大尺寸。然而，当请求填充时，可被填充的区域大小会受限于当前 GameCanvas 的尺寸，一个存在的 Ticker、Command 等都会影响到 GameCanvas 的尺寸。GameCanvas 的当前尺寸可以通过调用 getWidth 和 getHeight 方法获得。

一个游戏可以提供自己的线程来运行游戏循环。一个典型的循环将检查输入，实现游戏逻辑，然后渲染更新后的用户界面。以下代码演示了一个典型的游戏循环的结构：

```
//从后备屏幕缓冲获得 Graphics 对象
Graphics g = getGraphics();
while (true) {
        //如果有需要，检查用户输入并更新位置
        int keyState = getKeyStates();
        if ((keyState & LEFT_PRESSED) != 0) {
        sprite.move(-1, 0);
}
        else if ((keyState & RIGHT_PRESSED) != 0) {
        sprite.move(1, 0);
}
//将背景清除成白色
g.setColor(0xFFFFFF);
g.fillRect(0,0,getWidth(), getHeight());
//绘制 Sprite（精灵）
sprite.paint(g);
//输出后备缓冲区的内容
flushGraphics();
}
```

6.2.1 绘图

要创建一个新的 GameCanvas 实例，只能通过继承并调用父类的构造函数：

- protectedGameCanvas(booleansuppressKeyEvents)

这将使为 GameCanvas 准备的一个新的缓冲区也被创建并在初始化时被填充为白色。

为了在 GameCanvas 上绘图，首先要获得 Graphics 对象来渲染 GameCanvas。

- protectedGraphicsgetGraphics()

返回的 Graphics 对象将用于渲染属于这个 GameCanvas 的后备屏幕缓冲区（off-screen buffer），但是渲染结果不会立刻显示出来，直到调用 flushGraphics()方法，输出缓冲区的内容也不会改变后备屏幕缓冲区的内容，即输出操作不会清除缓冲区的像素。

每次调用这个方法时，都会创建一个新的 Graphics 对象；对于每个 GameCanvas 实例，获得的多个 Graphics 对象都将渲染同一个后备屏幕缓冲区。因此，有必要在游戏启动前获得并存储 Graphics 对象，以便游戏运行时能反复使用。

刚创建的 Graphics 对象有以下属性：

- 渲染目标是这个 GameCanvas 的缓冲区；
- 渲染区域覆盖整个缓冲区；
- 当前颜色是黑色（black）；
- 字体和调用 Font.getDefaultFont()返回的缺省字体相同；

- 绘图模式为 SOLID；
- 坐标系统的原点定位在缓冲区的左上角。

在完成了绘图操作后，可以使用 flushGraphics()方法将后备屏幕缓冲区的内容输出到显示屏上，输出区域的大小与 GameCanvas 的大小相同，输出操作不会改变后备屏幕缓冲区的内容。这个方法直到输出操作完成后才返回，因此，当这个方法返回时，应用程序可以立刻对缓冲区进行下一帧的渲染。

如果 GameCanvas 当前没有被显示，或者系统忙而不能执行输出请求，这个方法将不进行任何操作就立刻返回。

6.2.2 键盘

如果需要，开发者可以随时调用 getKeyStates 方法来查询键的状态。getKeyStates()获取游戏的物理键状态，返回值的每个比特位都表示设备上的一个特定的键。如果一个键对应的比特位的值为 1，表示该键当前被按下，或者自上次调用此方法后到现在，至少被按下过一次；如果一个键对应的比特位的值为 0，表示该键当前未被按下，并且自上次调用此方法后到现在从未被按下过。这种"闭锁行为（latching behavior）"保证了不管循环有多慢一个快速的按键和释放总是能够在游戏循环中被捕获。下面是获取游戏按键的示例：

```
//获得键的状态并存储
int keyState = getKeyStates();
if ((keyState & LEFT_KEY) != 0) {
    positionX--;
}
else if ((keyState & RIGHT_KEY) != 0) {
    positionX++;
}
```

调用这个方法的副作用是不能及时清除过期的状态。在一个 getKeyStates 调用后如果紧接着又一个调用，键的当前状态将取决于系统是否已经清除了上一次调用后的结果。

某些设备可能无法直接访问键盘硬件，因此，这个方法可能是通过监视键的按下和释放事件来实现的，这会导致 getKeyStates 滞后于当前物理键的状态，延时取决于每个设备的性能。某些设备还可能没有探测多个键同时按下的能力。

请注意，除非 GameCanvas 当前可见（通过调用 Displayable.isShown()方法），否则此方法返回 0。一旦 GameCanvas 变为可见，将初始化所有键为未按下状态 0。

6.3 Sprite 的使用

Sprite 是一个基本的可视元素，可以用存储在图像中的一帧或多帧来渲染它，轮流显示不同的帧可以令 Sprite 实现动画。翻转和旋转等几种变换方式也能应用于 Sprite 使之外观改变。作为 Layer 的子类，Sprite 的位置可以改变，并且还能设置其可视与否。

6.3.1　Sprite 帧

用于渲染 Sprite 的原始帧由一个单独的 Image 对象提供，此 Image 可以是可变的，也可以是不可变的。如果使用多帧，图像将按照指定的宽度和高度被切割成一系列相同大小的帧。如图 6-1 所示，同一序列的帧可以以不同的排列存储，这取决于是否方便游戏开发者开发。

每一帧都被赋予一个唯一的索引号。左上角的帧被赋予索引号 0。余下的帧按照行的顺序索引号依次递增（索引号从第一行开始，接着是第二行，以此类推）。getRawFrameCount()方法返回所有原始帧的总数。

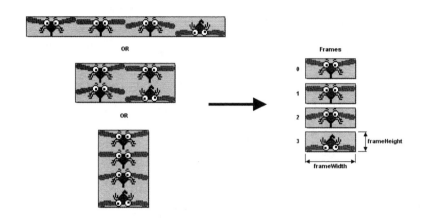

图 6-1　Sprite 示意图

6.3.2　帧序列

Sprite 的帧序列定义了帧以什么样的顺序来显示。缺省的帧序列就是所有可用帧的顺序排列，因此，帧序列和对应的帧的索引号是一致的。这表示缺省的帧序列的长度和所有原始帧的总数是相等的。例如，如果一个 Sprite 有 4 帧，则缺省的帧序列为 {0, 1, 2, 3}，如图 6-2 所示。可以使用 setFrameSequence(int[] sequence)来为 Sprite 设置帧序列。当调用此方法时，将会复制 sequence 数组，因此，随后对参数 sequence 数组进行的任何更改均不会影响 Sprite 的帧序列；传入 null 将使 Sprite 的帧序列重置为缺省值。开发者必须在帧序列中手动切换当前帧，可以调用 setFrame(int)、prevFrame()或者 nextFrame()方法来完成。

注意，以上这些方法是针对帧序列操作，而不是针对帧的索引操作。如果使用缺省的帧序列，那么帧序列的索引和帧的索引是可互换的。

图 6-2　Sprite 帧序列示意图-1

　　如果愿意，可以为 Sprite 定义任意的帧序列。帧序列必须至少包含一个元素，并且每个元素都必须有一个有效的帧的索引号。通过定义新的帧序列，开发者可以方便地以任意想要的顺序来显示 Sprite 的帧，即帧可以重复、忽略或者以相反的顺序显示。

　　例如，图 6-3 显示了一个特定的序列如何被用于动画显示一只蚊子。帧序列被设计为蚊子振动翅膀 3 次，然后在下次循环前暂停一会儿。

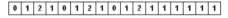

图 6-3　Sprite 帧序列示意图-2

每次调用 nextFrame()方法就会更新显示，动画效果如图 6-4 所示。

图 6-4　Sprite 帧序列示意图-3

要创建一个静态的 Sprite，可以调用 public Sprite(Image image)方法通过提供的图像创建一个新的 Sprite。要创建一个动态的 Sprite，则必须使用 public Sprite(Image image,int frameWidth, int frameHeight)。

帧的大小由 frameWidth 和 frameHeight 指定，所有帧的大小必须相等。可以在图像中水平、竖直或以方格形式排列帧。源图像的宽度必须是帧宽度的整数倍，高度也必须是帧高度的整数倍。如果 image 的宽度或高度不是 frameWidth 或 frameHeight 的整数倍，将会抛出 IllegalArgumentException 异常。

6.3.3　ReferencePixel

作为 Layer 的一个子类，Sprite 继承了很多方法来设置和获取位置，如 setPosition(x,y)、getX()和 getY()。这些方法定义的位置均以 Sprite 视图区域的左上角像素点为依据。然而，在某些情况下，根据 Sprite 的其他像素点来定位 Sprite 更加方便，尤其是在 Sprite 上应用某些转换时。

因此，Sprite 引入一个参考像素点（reference pixel）的概念。参考像素点通过使用 defineReferencePixel(x,y)方法指定其在 Sprite 未经变换的帧内的某一点来定义。缺省的参考像素点定义在帧的(0,0)像素点。如果有必要，参考像素点也可以定义在帧区域以外。

在图 6-5 所示例子中，参考像素点被定义在猴子悬挂的手上。

图 6-5　参考点-1

getRefPixelX()和 getRefPixelY()方法可用于查询参考像素点在绘图坐标系统中的位置。开发者也可以调用 setRefPixelPosition(x,y)方法来定位 Sprite，使得参考像素点定义在绘图坐标系统中的指定位置。这些方法会自动地适应任何应用在 Sprite 上的变换。

在图 6-6 所示的例子中，参考像素点被定位在树枝末端的一点，Sprite 的位置也改变了，使得参考像素点定位在这一点上，猴子看起来像挂在树枝上一样。

图 6-6　参考点-2

6.3.4　Sprite 的变换

几种变换可应用于 Sprite。可用的变换包括旋转几个 90°再加上镜像（沿垂直轴）。Sprite 的变换通过调用 setTransform(transform)方法实现，如图 6-7 所示。

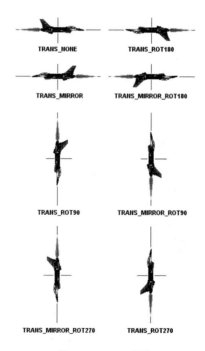

图 6-7　Sprite 变换

当应用一个变换时，Sprite 被自动重新定位，使得参考像素点在绘图坐标系统中看起来是静止的。因此，参考像素点即为变换操作的中心点。因为参考像素点并未移动，getRefPixelX()和 getRefPixelY()方法返回的值仍不变。但是，getX()和 getY()方法可能改变，以便反映出 Sprite 左上角位置的移动。

再次回到猴子的例子上来，当应用一个 90°旋转后，参考像素点的位置仍然在(48, 22)，因此使得猴子像是沿着树枝飘着，如图 6-8 所示。

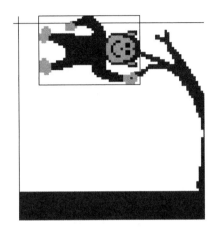

图 6-8　Sprite 像素参考点

由于某些变换涉及到 90°或 270°旋转，其使用结果可能导致 Sprite 的宽度和高度互换。因此，调用 Layer.getWidth()和 Layer.getHeight()方法的返回值可能改变。

6.3.5　绘制 Sprite

可以在任何时候通过调用 paint(Graphics)方法来绘制 Sprite。根据 Sprite 当前的状态信息（如位置、帧、可视与否等）Sprite 将被绘制在 Graphics 对象上。擦除 Sprite 通常是 Sprite 以外的类的责任。

厂商可以使用任何希望使用的技术（如硬件加速可以用于所有 Sprite 或特定大小的 Sprite，或者根本不使用硬件加速）来实现 Sprite。对一些平台而言，特定大小的 Sprite 可能比其他大小的 Sprite 更高效，厂商可以选择提供给开发者与设备相关的这些特性。

6.3.6　碰撞检测

Sprite 非常适合移动的物体，如游戏主角、敌人等等。在游戏中，可以使用 Sprite 提供的碰撞检测功能来简化游戏逻辑。

使用 defineCollisionRectangle()方法可定义用于碰撞检测的 Sprite 的矩形区域,此指定的矩形是相对于未经变换的 Sprite 的左上角,该区域将用于检测碰撞。对于像素级的碰撞检测,仅仅在这个碰撞检测区内部的像素点会被检查。通常 Sprite 的碰撞检测区定位在(0,0),并与 Sprite 尺寸相同。碰撞检测区也可以指定为大于或小于缺省的碰撞检测区,如果大于缺省的碰撞检测区,在 Sprite 之外的像素在像素级的碰撞检测时会被认为是透明的。

要判断两个 Sprite 是否碰撞,或者与其他 Layer 是否碰撞,可以使用 collidesWith()方法。如果使用像素级检测,仅当非透明像素重叠时,碰撞才被检测到。即第一个 Sprite 中的非透明像素和第二个 Sprite 中的非透明像素重叠时,碰撞才被检测到。仅仅那些包含在 Sprite 的碰撞检测区内的像素会被检测。

如果不使用像素级检测,这个方法就简单地检查两个 Sprite 的碰撞检测区矩形是否有重合。如果对 Sprite 应用了变换,还会进行相应的处理。注意,只有两个 Sprite 都可见时,才能检测碰撞。

6.4 Layer 的使用

Layer 是一个抽象类,表示游戏中的一个可视元素。上节中讲述的 Sprite 就是 Layer 的一种。每个 Layer 都有位置(取决于它的左上角在容器中的位置):宽度、高度和可视与否。Layer 的子类必须实现一个 paint(Graphics)方法,使得它们能够被渲染。如果该 Layer 可见,就从它的左上角开始渲染,其当前坐标(x,y)是相对于原始的 Graphics 对象。当渲染 Layer 时,应用程序可以使用剪辑和坐标变换来控制并限制渲染的区域。实现此方法的子类有责任检查 Layer 是否可见,如果不可见,这个方法不执行任何操作。此外,调用此方法不应该改变 Graphics 对象的属性(剪辑区域、坐标变换、绘图颜色等)。

Layer 的位置坐标(x,y)通常都是相对于 Graphics 对象的坐标系统,该对象通过 Layer 的 paint()方法传递。这个坐标系统被称为绘图坐标系统。一个 Layer 的初始位置是(0,0)。

6.4.1 TiledLayer

TiledLayer 由一系列单元格组成,单元格可被一组贴图填充。这个类允许不必使用特别大的图像来创建大的虚拟层,该技术在 2D 游戏中被广泛用于创建特别大的可卷动的背景。

1. 贴图(Tiles)

贴图用于填充 TiledLayer 的单元格,由一个单一的可变或不可变的 Image 对象提供。图像被切割成一系列相同大小的贴图,贴图大小随 Image 一同指定。如图 6-9 所示,相同的一系列贴图可以以不同的方式存储,取决于游戏开发者方便与否。

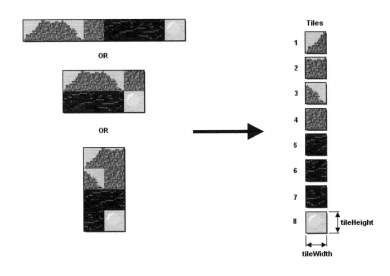

图 6-9 贴图

每个贴图都被赋予一个唯一的索引号。位于图像最左上角的贴图被赋予索引号 1。剩下的贴图按照一行一行的顺序（首先是第一行，然后是第二行，以此类推）索引号依次递增。这些贴图被视为静态贴图（static tiles），因为贴图和图像内容有固定的联系。

当实例化一个 TiledLayer 时，一组静态贴图就被创建了，也可以在任何时候调用 setStaticTileSet(javax.microedition.lcdui.Image, int, int)方法来更新它们。

除了静态贴图外，开发者同样能够定义一系列动态贴图（animated tiles）。一个动态贴图就是一个虚拟的贴图，它与一个静态贴图动态地联系在一起，一个动态贴图的外观就是当时与之联系的静态贴图。

动态贴图允许开发者非常容易地改变一组单元格的外观。对于用动态贴图填充的单元格而言，改变它们的外观仅仅需要简单地改变与动态贴图关联的静态贴图即可。此技术对于动画显示大的重复性区域非常有用，因为不需要显式地改变大量单元格的内容。

动态贴图可以通过调用 createAnimatedTile(int)方法来创建，该方法返回一个索引号，用于标记新创建的动态贴图。动态贴图的索引号总是负数，并且也是连续的，起始值为-1。一旦被创建，与之关联的静态贴图可以通过调用 setAnimatedTile(int, int)方法来改变。

2．单元格（Cells）

TiledLayer 由相同大小的单元格组成，每行和每列的单元格数目在构造方法中指定，实际大小取决于贴图的大小。

每个单元格的内容由贴图索引号指定，一个正的贴图索引号代表一个静态贴图，一个负的贴图索引号代表一个动态贴图。索引号为 0 的贴图表示该单元格为空，为空的单元格是完全透明的，并且不会被 TiledLayer 绘制任何内容。缺省的，所有单元格都包含索引号为 0 的贴图。

可以通过调用 setCell(int, int, int)和 fillCells(int, int, int, int, int)方法来改变单元格的内容。很多单元格可以包含同一个贴图，但一个单元格仅能包含一个贴图。图 6-10 所示的例子演示了如何使用 TiledLayer 来创建一个简单的背景。

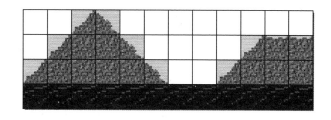

图 6-10　单元格-1

在这个例子中，水的区域由动态贴图来填充，索引号为-1，该动态贴图在初始化时与一个索引号为 5 的静态贴图关联。可以简单地通过调用 setAnimatedTile(-1, 7)方法来改变与之联系的静态贴图，从而实现整个水区域的动画效果，如图 6-11 所示。

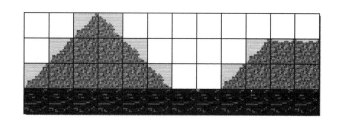

图 6-11　单元格-2

可以通过手动调用 paint()方法来渲染一个 TiledLayer；也可以使用 LayerManager 对象自动渲染它。绘图方法将尝试渲染在 Graphics 对象的剪裁区域内的整个 TiledLayer，从 TiledLayer 的左上角开始渲染，该点的当前坐标(x,y)相对于 Graphics 对象的原点。渲染区域也可以通过设置 Graphics 对象的剪裁区域来控制。

6.4.2　LayerManager

LayerManager 管理一系列的 Layer，它简化了渲染每个 Layer 的过程，每个添加的 Layer 都将在正确的区域并以正确的顺序被渲染。

LayerManager 维护一个顺序列表，以便管理如何追加、插入和删除 Layer。一个 Layer 的索引号关联了它的 Z 轴位置（z-order），索引号为 0 的 Layer 最接近用户，索引号越大的 Layer 离用户越远。索引号永远是连续的，如果一个 Layer 被删除，后面的 Layer 的索引号都将调整使得索引号保持连续。

LayerManager 类提供一些用于控制游戏中如何在屏幕上渲染 Layer 的功能。

可视窗口（view window）控制着可视区域及其在 LayerManager 的坐标系统中的位置，改变可视窗口的位置可以实现上下或左右滚动屏幕的效果。例如，如果想向右移动，只需简单地将可视窗口的位置右移。可视窗口的大小决定了用户的可视范围，通常它应该适合设备的屏幕大小。

在图 6-12 所示的例子中，可视窗口被设置为 85×85 像素大小，并定位在 LayerManager 的坐标系统的(52,11)点。每个 Layer 的位置都是相对于 LayerManager 的原点。

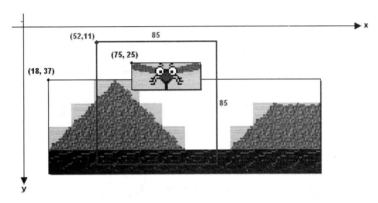

图 6-12　图层管理-1

paint(Graphics, int, int)方法包含一个(x,y)坐标，控制可视窗口在屏幕中的显示位置。改变参数不会改变可视窗口的内容，仅仅简单地改变可视窗口在屏幕中被绘制的位置。注意到这个位置是相对于 Graphics 对象的原点而言的，因此它服从 Graphics 对象的变换属性。

例如，如果一个游戏在屏幕的最顶端显示分数，可视窗口可能在(17,17)点被渲染，确保有足够的空间来显示分数，如图 6-13 所示。

图 6-13　图层管理-2

我们使用 append()方法向这个 LayerManager 添加一个 Layer，Layer 将被添加到现有 Layer 列表的末尾，即拥有最大的索引号（离用户最远）。如果此 Layer 已存在，在添加前将首先被删除。

insert()方法与 append()方法的区别在于可以指定 Layer 的索引号。如果此 Layer 已存在，将在添加前首先被删除。

获得指定位置的 Layer 可以调用 getLayerAt(int index)方法。渲染 LayerManager 的 paint() 方法将以降序的顺序来渲染每一个 Layer，以保证实现正确的 Z 轴次序。完全在可视窗口之外的 Layer 将不被渲染。

paint 方法的另外两个参数决定了 LayerManager 的可视窗口相对于 Graphics 对象的原点在何处渲染。例如，一个游戏可能在屏幕上方显示分数，因此游戏的 Layer 就必须在这个区域下面，可视窗口可能在点(0,20)处开始渲染。此位置相对于 Graphics 对象的原点，因此 Graphics 对象的坐标转换模式也会影响可视窗口在屏幕上渲染的位置。

Graphics 对象的剪裁区域被设置为与位于(x,y)处的可视窗口的区域一致。LayerManager 将转换 Graphics 对象的坐标，使得点(x,y)与可视窗口在 LayerManager 中的坐标系统的位置一致。然后，Layer 以一定的次序被渲染。在方法返回前，Graphics 对象的坐标转换模式和剪裁区域将重置为原先的值。

渲染会自动适应 Graphics 对象的剪裁区域和变换方式。这样，如果剪裁区域不够大，可视窗口将仅有部分被渲染。

为了提升速度，paint 方法可能忽略不可见的 Layer，或者全部在 Graphics 对象剪裁区域以外的 Layer。在调用 Layer 的 paint()方法前，Graphics 对象并不会重置为一个确定的状态。剪裁区域也可能大于 Layer 的区域，此时 Layer 必须自己保证渲染操作在其范围内进行。

6.5 一个示例

本节给出一个示例游戏的核心代码，这是一个著名的潜艇游戏的手机版本，也是一个良好的游戏入门范本，其中涉及到精灵的使用、Tiled 的使用以及碰撞检测等运动类基于 Tiled 的 2D 游戏常见的问题。希望通过对它的学习能给你一些启示。为了配合本章 API 的讲解，我们省略了大部分的游戏周边代码，给出的仅仅是游戏的 GameCanvas 子类，这个类同时包含了游戏的主循环线程。

如你所见，这不是一个产品级的游戏，它只教会你基本的内容，而不是全部。同样这里没有包括什么优化。图 6-14 为游戏的截图。另外本游戏所用的资源大多来源于网络，代码仅供非商业用途的学习参考。

请着重注意代码中的以下方法：
paintCanvas 方法用于渲染；
run 方法用于游戏的主循环；

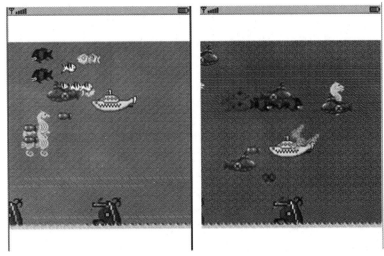

图 6-14　潜艇游戏实例

关于输入捕获的一点说明是，这个游戏并没有屏蔽键盘事件，它混合使用了主动轮询用于潜艇运动，而开火则采用捕获方式。

```
import javax.microedition.lcdui.*;
import javax.microedition.lcdui.game.GameCanvas;
import javax.microedition.lcdui.game.*;
import javax.microedition.lcdui.game.LayerManager;
import java.util.*;
public class SubCanvas extends GameCanvas implements Runnable, CommandListener {
    /**
     *
     * @uml.property name="subMIDlet"
     * @uml.associationEnd multiplicity="(0 1)"
     */
    private Controller controller;
    private Graphics graphics;
    private Thread thread;
    private boolean threadAlive = false;
    private Command startCommand;
    private Command exitCommand;
    private Command pauseCommand;
    //图层数据
    private LayerManager layerManager;
    private TiledLayer layerSeaback;
    private Sprite spriteMap;
    public final int GAME_INIT = 0; //游戏初始状态
```

```java
public final int GAME_RUN = 1; //游戏运行状态
public final int GAME_OVER = 4; //游戏结束状态
public final int GAME_PAUSE = 5;
public final int GAME_SUSPEND = 9; //暂停状态
public boolean COMMAND_ADD_FLAG = false; //是否已经添加 Command 标识
public static final int TROOP_PLAYER = 0; //敌我标识
public static final int TROOP_ENEMY = 1;
public static int PLAYER_LEVEL = 1; //当前玩家水平
public static int ENEMY_MAX = PLAYER_LEVEL * 10; //最大敌人数量
public static int ENEMY_CURRENT = 0; //当前敌人数量
public static int ENEMY_CURRENT_LIMIT = 0; //当前敌人数量限制
protected int TRIGGER_COUNT = 0; //拖延标识，避免敌人新潜艇同一时刻全部产生
protected int TICK_COUNT = 0;
public static int mainWidth; //屏幕宽度
public static int mainHeight; //屏幕高度
public int gameState; //游戏状态
public final static int TILE_WIDTH = 10; //背景单元宽度 10px
public final static int TILE_HEIGHT = 10; //背景单元高度 10px
/**
 * 动画变化频率(200ms) <code>MILLIS_PER_TICK</code> 的注释
 */
public final static int MILLIS_PER_TICK = 200;
private final static int WIDTH_IN_TILES = 45; //游戏域宽度（以单元宽度计算）16 .. N
private final static int HEIGHT_IN_TILES = 24; //游戏域高度（以单元高度计算）
private final static int NUM_DENSITY_LAYERS = 4; //海面密度（背景图层）
private int[] rotations = { Sprite.TRANS_NONE, Sprite.TRANS_MIRROR,
        Sprite.TRANS_MIRROR_ROT90, Sprite.TRANS_MIRROR_ROT180,
        Sprite.TRANS_MIRROR_ROT270, Sprite.TRANS_ROT90,
        Sprite.TRANS_ROT180, Sprite.TRANS_ROT270 };
/**
 * 游戏背景宽度（以像素计算：宽度单元数 * 宽度单元像素）
 * <code>WORLD_WIDTH</code> 的注释
 */
public final static int WORLD_WIDTH = WIDTH_IN_TILES * TILE_WIDTH;

/**
 * 游戏背景高度（以像素计算：高度单元数 * 高度单元像素）
 * <code>WORLD_HEIGHTH</code> 的注释
 */
public final static int WORLD_HEIGHT = HEIGHT_IN_TILES * TILE_HEIGHT;
```

```java
private final static int NUM_DENSITY_LAYER_TILES = 4; //每一个密度层的 TILE 单元数
private final static int FRACT_DENSITY_LAYER_ANIMATE = 20;
private int SEABACK_DENSITY; //游戏初始海水密度为 0
//private final Vector oceanLayersVector = new Vector();
private Vector fishCollectionVector = new Vector();
public Vector enemyCollectionVector = new Vector();
public Vector tinfishCollectionVector = new Vector();
/**
 *
 * @uml.property name="mySub"
 * @uml.associationEnd multiplicity="(0 1)"
 */
private Sub mySub = null;
private Runtime rt = null;
//初始化为不使用窗口区域视野
private boolean userViewWindow = false;
//创建不稳定的动画线程
private volatile Thread animationThread = null;
//LayerManager 的偏移坐标
private int xViewWindow;
private int yViewWindow;
private int wViewWindow;
private int hViewWindow;
public SubCanvas(Controller controller) {
    //不屏蔽键盘事件（潜艇运动采用主动轮询，而开火则采用捕获方式）
    super(false);
    this.controller = controller;
    this.graphics = getGraphics();
    this.layerManager = new LayerManager();
    //决定图层显示方式
    init();
    //画布构造即建立玩家潜艇
    //初始位置为屏幕的(1/3, 1/3)位置
    mySub = new Sub(this, SubMIDlet.createImage("/res/sub.png"),
            mainWidth / 3, mainHeight / 3, layerManager);
    //监测运行时，以便及时运行垃圾回收
    rt = Runtime.getRuntime();
    startCommand = new Command("Start", Command.OK, 1);
    pauseCommand = new Command("Pause", Command.OK, 1);
    exitCommand = new Command("Exit", Command.EXIT, 2);
```

```java
        //初始化其他类及图层
        //初始化游戏状态
        this.gameState = this.GAME_INIT; //游戏处于demo 画面状态
        //启动应用程序
        threadAlive = true;
        thread = new Thread(this);
        thread.start();
    }

    /**
     * 初始化地图数据和地图窗口显示方式
     */
    private void init() {
        //清理数据
        this.clearData();
        mainWidth = getWidth();
        mainHeight = getHeight();
        //判断是否使用预览模式窗口
        //根据显示设备,设置合适的最大区和显示视野
        this.xViewWindow = 0;
        if (WORLD_WIDTH > mainWidth) {
            //现有设备不能容纳所有游戏区域
            userViewWindow = true;
            this.wViewWindow = mainWidth;
        } else {
            //现有设备可以容纳所有游戏区域
            this.wViewWindow = WORLD_WIDTH;
        }
        this.yViewWindow = 0;
        if (WORLD_HEIGHT > mainHeight) {
            userViewWindow = true;
            this.hViewWindow = mainHeight;
        } else {
            this.hViewWindow = WORLD_HEIGHT;
        }
        //设定图层显示方式
        if (userViewWindow) {
            this.layerManager.setViewWindow(xViewWindow, yViewWindow,
                    wViewWindow, hViewWindow);
        }
```

```java
}

protected void clearData() {
    PLAYER_LEVEL = 1;
    ENEMY_MAX = PLAYER_LEVEL * 10;
    ENEMY_CURRENT = 0;
    TRIGGER_COUNT = 0;
    SEABACK_DENSITY = 0;
    ENEMY_CURRENT_LIMIT = PLAYER_LEVEL * 2;
}

/**
 * 程序作为线程,每50ms运行刷新一次
 */
public void run() {
    //利用条件驱动线程
    while (threadAlive) {
        try {
            Thread.sleep(25);
        } catch (InterruptedException e) {
            e.printStackTrace();
        }
        //分离对玩家潜艇和普通物体的响应速度(一倍)
        if (gameState == GAME_RUN) {
            mySub.movePosition(getKeyStates());
        }
        if ((TICK_COUNT % 2) == 0) {
            //重画事件
            this.paintCanvas(graphics);
            this.tick();
        }
        TICK_COUNT++;
    }
}

protected synchronized void keyPressed(int keyCode) {
    int action = getGameAction(keyCode);
    if (action == Canvas.FIRE && gameState == GAME_RUN) {
        //玩家潜艇开火
        if (mySub != null) {
```

```
                    mySub.fire();
            }
        }
    }

    /**
     * 秒触发器
     */
    public synchronized void tick() {
        //主动查询状态
        int keyState = getKeyStates();
        if (gameState != GAME_OVER) {
            //执行鱼雷图形运动
            //执行鱼雷触发
            this.tickTinfish();
            if (gameState == GAME_RUN) {
                //替代 Canvas 中的 KeyPressed()捕获事件
                //mySub.movePosition(keyState);
                //创建并执行敌人潜艇行为
                this.tickSub();
                this.tickEnemySub();
                if (ENEMY_CURRENT == 0 && ENEMY_MAX == 0) {
                    gameState = GAME_SUSPEND;
                }
            }
        }
    }

    /**
     * 鱼雷图形运动
     */
    private void tickFishes() {
        for (int i = 0; i < fishCollectionVector.size(); i++) {
            FishCollection collection = (FishCollection) fishCollectionVector
                    .elementAt(i);
            collection.tick();
        }
    }

    /**
```

```
 * 执行鱼雷触发
 */
protected void tickTinfish() {
    //鱼雷图形运动
    //如果某个鱼雷元素已经结束生命周期，则置 null，并从数组中删除
    for (int j = 0; j < tinfishCollectionVector.size(); j++) {
        Tinfish tinfish = (Tinfish) tinfishCollectionVector.elementAt(j);
        //当生命周期结束
        if (!tinfish.isLifeState()) {
            tinfishCollectionVector.removeElementAt(j);
            this.layerManager.remove(tinfish);
            tinfish = null;
        } else {
            tinfish.tick();
        }
    }
}

/**
 * 玩家潜艇运动及生命状态
 */
private void tickSub() {
    if (!mySub.isLifeState()) {
        gameState = GAME_OVER;
    }
}

/**
 * 创建并执行敌人潜艇的运行操作
 */
protected void tickEnemySub() {
    //当敌人剩余最大数量大于 0，并且敌人当前数量小于并行敌人上限时
    //可以添加新的敌人潜艇
    if (ENEMY_MAX >= 0 && ENEMY_CURRENT <= ENEMY_CURRENT_LIMIT
            && ENEMY_CURRENT < 10) {
        int iLeft = ENEMY_MAX - ENEMY_CURRENT;
        //当剩余敌人量(最大量 - 当前量)大于 0 的时候
        if (iLeft > 0) {
            int n = SubMIDlet.createRandom(iLeft) + 1;
            Image image = SubMIDlet.createImage("/res/enemysub_f.png");
```

```java
            int xPosition = 0;
            int yPosition = (SubMIDlet.createRandom(5) * WORLD_HEIGHT) / 5;
            for (int i = 0; i < n; i++) {
                //拖延标识，避免敌人新潜艇同一时刻全部产生
                if (TRIGGER_COUNT >= 20) {
                    yPosition = (WORLD_HEIGHT * (i % 5)) / 5;
                    if (i % 2 == 0) {
                        xPosition = 0;
                    } else {
                        xPosition = WORLD_WIDTH - image.getWidth();
                    }
                    //创建一艘敌人潜艇，同时更新监听数组
                    EnemySub enemySub = new EnemySub(this, image,
                            xPosition, yPosition, PLAYER_LEVEL);
                    ENEMY_CURRENT++;
                    layerManager.insert(enemySub, 0);
                    this.enemyCollectionVector.addElement(enemySub);
                    TRIGGER_COUNT = 0;
                } else {
                    TRIGGER_COUNT++;
                }
            }
            image = null;
        }
}
//对所有已经存在的敌人潜艇进行 tick 触发
Image imageDestroyed = null;
for (int j = 0; j < enemyCollectionVector.size(); j++) {
    EnemySub enemySub = (EnemySub) enemyCollectionVector.elementAt(j);
    int iCount = 0;
    //当生命周期结束
    if (!enemySub.isLifeState()) {
        imageDestroyed = SubMIDlet.createImage("/res/enemysub_die.png");
        enemySub.setImage(imageDestroyed, imageDestroyed.getWidth(),
                imageDestroyed.getHeight());
        if (enemySub.getVx() >= 0) {
            enemySub.setTransform(Sprite.TRANS_NONE);
        } else {
            enemySub.setTransform(Sprite.TRANS_MIRROR);
        }
```

```
                enemyCollectionVector.removeElementAt(j);
                //消灭一艘敌人潜艇，同时更新监听数组
                SpriteChanged spriteChanged = new SpriteChanged(enemySub,
                        layerManager);
                spriteChanged.start();
                spriteChanged = null;
                enemySub = null;
                ENEMY_CURRENT--;
                ENEMY_MAX--;
            } else {
                enemySub.tick();
            }
        }
        imageDestroyed = null;
    }

    /**
     * 画布重画事件，替代 Canvas 中的 paint()事件，根据不同的游戏状态（gameState）画出游戏画面
     * 本方法由 thread 每次重新启动，最后执行 flushGraphics()重画缓冲区
     */
    public synchronized void paintCanvas(Graphics g) {
        if (gameState == GAME_INIT) {
            //游戏第一次启动
            //设置为全屏模式并清屏
            this.setFullScreenMode(true);
            g.setColor(255, 255, 255);
            g.fillRect(0, 0, mainWidth, mainHeight);
            if (!COMMAND_ADD_FLAG) {
                //添加响应命令及监听器
                this.addCommand(pauseCommand);
                this.addCommand(exitCommand);
                this.setCommandListener(this);
                COMMAND_ADD_FLAG = true;
            }
            if (fishCollectionVector != null) {
                fishCollectionVector = null;
                fishCollectionVector = new Vector();
            }
            if (this.layerManager != null) {
                this.layerManager = null;
```

```java
            this.layerManager = new LayerManager();
            if (userViewWindow) {
                    this.layerManager.setViewWindow(xViewWindow, yViewWindow,
                            wViewWindow, hViewWindow);
            }
        }
        if (mySub != null) {
            mySub = null;
            mySub = new Sub(this, SubMIDlet.createImage("/res/sub.png"),
                    getWidth() / 3, getHeight() / 3, layerManager);
        }
        //创建背景图层
        this.createSandBackground();
        this.createSunkenBoat();
        this.createFishCollection(0, FishCollection.NUM_FISH_TYPES);
        this.createMysub();
        this.createSeaBackground();
        mySub.setPosition(getWidth() / 3, getHeight() / 3);
        gameState = GAME_RUN;
    } else if (gameState == GAME_RUN) {
        //游戏处于运行状态
        //提供游戏运行标识 Flag，保证对用户操作的响应和"敌人"的运行动作只在运行时生效

        //在性能过耗（可用内存不到当前内存总量的 4/5）时，进行垃圾回收 GC
        if (rt.freeMemory() < (rt.totalMemory() * 4 / 5)) {
            rt.gc();
        }
    } else if (gameState == GAME_SUSPEND) {
        //下一轮游戏
        //更新数据
        mySub.setHpLife(15);
        mySub.setPosition(mainWidth / 3, mainHeight / 3);
        if (PLAYER_LEVEL >= 4) {
            gameState = GAME_OVER;
        } else {
            PLAYER_LEVEL++;
            controller.EventHandler(Controller.EVENT_NEXTROUND);
            //目前游戏只设计了四级关卡
            ENEMY_MAX = PLAYER_LEVEL * 10;
            ENEMY_CURRENT = 0;
            TRIGGER_COUNT = 0;
```

```java
                    TICK_COUNT = 0;
                    ENEMY_CURRENT_LIMIT = PLAYER_LEVEL * 2;
                    Layer layer = null;
                    //暂时删除 LayerManager 中所有文件，清空鱼雷与敌人潜艇数据
                    //为更新图层做准备
                    this.tinfishCollectionVector.removeAllElements();
                    this.enemyCollectionVector.removeAllElements();
                    this.fishCollectionVector.removeAllElements();
                    //for(int i = 0; i < layerManager.getSize(); i++){
                    //layer = layerManager.getLayerAt(i);
                    //layerManager.remove(layer);
                    //}
                    layer = null;
                    //更新海底图层数据
                    this.SEABACK_DENSITY = (SEABACK_DENSITY + 1) % 4;
                    gameState = GAME_INIT;
                    this.unActive();
                }
            } else if (gameState == GAME_OVER) {
                //游戏结束
                threadAlive = false;
                controller.EventHandler(Controller.EVENT_MENU_GAMEOVER);
                gameState = this.GAME_INIT;
            }
            //在缓冲区重画
            this.layerManager.paint(g, 0, (getHeight() - WORLD_HEIGHT) / 2);
            this.flushGraphics();
        }

        public void commandAction(Command command, Displayable display) {
            if (command == startCommand) {
                if (this.gameState == GAME_OVER) {
                    gameState = GAME_INIT;
                } else {
                    gameState = GAME_RUN;
                }
                this.removeCommand(this.startCommand);
                this.addCommand(pauseCommand);
            } else if (command == pauseCommand) {
                gameState = GAME_PAUSE;
                this.removeCommand(this.pauseCommand);
```

```java
                this.addCommand(startCommand);
        } else if (command == exitCommand) {
                gameState = GAME_OVER;
        }
}

/**
 * 创建海底沙地背景图层
 */
protected void createSandBackground() {
    Image bottomTitles = SubMIDlet.createImage("/res/bottom.png");
    //将图片 bottomTitles 切成指定大小(TILE_WIDTH, TILE_HEIGHT)
    //创建一个指定维数(1, WIDTH_IN_TILES)的背景数组
    TiledLayer layer = new TiledLayer(WIDTH_IN_TILES, 1, bottomTitles,
            TILE_WIDTH, TILE_HEIGHT);
    for (int column = 0; column < WIDTH_IN_TILES; column++) {
        //将海底图层数组中的每个小格用原始图片的第 i 块来填充
        int i = SubMIDlet.createRandom(NUM_DENSITY_LAYER_TILES) + 1;
        layer.setCell(column, 0, i);
    }
    layer.setPosition(0, WORLD_HEIGHT - bottomTitles.getHeight());
    layerManager.append(layer);
    bottomTitles = null;
}

/**
 * 创建海底水层背景图片
 */
protected void createSeaBackground(){
    if(this.SEABACK_DENSITY >= 4){
        image = SubMIDlet.createImage("/res/densityLayer" +
                this.SEABACK_DENSITY + ".png");
        layerSeaback = new TiledLayer(WIDTH_IN_TILES, HEIGHT_IN_TILES,
                image, TILE_WIDTH, TILE_HEIGHT);
        for(int row = 0; row < HEIGHT_IN_TILES; row++){
            for(int column = 0; column < WIDTH_IN_TILES; column++){
            layerSeaback.setCell(column, row,
            SubMIDlet.createRandom(NUM_DENSITY_LAYER_TILES) + 1);
            }
        }
        layerSeaback.setPosition(0, 0);
```

```java
layerManager.append(layerSeaback);
image = null;
}
/**
 * 创建沉船图层
 */
protected void createSunkenBoat(){
    Image imageSunkenBoat = SubMIDlet.createImage("/res/sunkenBoat.png");
    //出现的沉船数量
    int numSunkenBoats = 1 + (WORLD_WIDTH / (3 * imageSunkenBoat.getWidth()));
    Sprite sunkenBoat;
    int bx = 0;
    for(int i = 0; i < numSunkenBoats; i++){
        sunkenBoat = new Sprite(imageSunkenBoat, imageSunkenBoat.getWidth(),
        imageSunkenBoat.getHeight());
        sunkenBoat.setTransform(this.rotations[SubMIDlet.createRandom(this.rotations.length)]);
        //随机定义沉船位置
        bx = (WORLD_WIDTH - imageSunkenBoat.getWidth()) / numSunkenBoats;
        bx = (i * bx) + SubMIDlet.createRandom(bx);
        sunkenBoat.setPosition(bx, WORLD_HEIGHT -
        imageSunkenBoat.getHeight());
        //添加图层
        this.layerManager.append(sunkenBoat);
        //mySub.addCollideable(sunkenBoat);
    }
    imageSunkenBoat = null;
}
/**
 * 创建玩家潜艇
 */
protected void createMysub(){
    this.layerManager.append(mySub);
}
/**
 * 创建鱼群背景
 *
 * @param startId
 * @param endId
 */
```

```java
                    protected void createFishCollection(int startId, int endId){
                        for(int id = startId; id < endId; id++){
                            int w = WORLD_WIDTH / 4;
                            int h = WORLD_HEIGHT / 4;
                            int x = SubMIDlet.createRandom(WORLD_WIDTH - w);
                            int y = SubMIDlet.createRandom(WORLD_HEIGHT - h);
                            int vx = FishCollection.randomVelocity(TILE_WIDTH / 4);
                            int vy = FishCollection.randomVelocity(TILE_HEIGHT / 4);
                            //初始化鱼类图层,同时把图层添加到图层管理器上
                            FishCollection fishCollection = new FishCollection(layerManager, id,
                            x, y, w, h, vx, vy, 0, 0, WORLD_WIDTH, WORLD_HEIGHT);
                            this.fishCollectionVector.addElement(fishCollection);
                        }
                    }
/**
 * 重新设置图层显示区域
 *
 * @param x
 *             玩家潜艇位置 x
 * @param y
 *             玩家潜艇位置 y
 * @param width
 *             玩家潜艇宽
 * @param height
 *             玩家潜艇高
 */
    public void adjustViewWindow(int xSub, int ySub, int width, int height) {
        if (this.userViewWindow)
        {
            xViewWindow = xSub + (width / 2) - (wViewWindow / 2);
            if (xViewWindow < 0)
            {
                xViewWindow = 0;
            }
            if (xViewWindow > (WORLD_WIDTH - wViewWindow))
            {
                xViewWindow = WORLD_WIDTH - wViewWindow;
            }
            yViewWindow = ySub + (height / 2) - (hViewWindow / 2);
            if (yViewWindow < 0)
```

```java
            {
                yViewWindow = 0;
            }
            if (yViewWindow > (WORLD_HEIGHT - hViewWindow))
            {
                yViewWindow = WORLD_HEIGHT - hViewWindow;
            }
            layerManager.setViewWindow(xViewWindow, yViewWindow, wViewWindow, hViewWindow);
    }
}
/**
 * 获取玩家潜艇
 *
 * @return 返回 mySub
 */
public Sub getMySub() {
    return mySub;
}
/**
 * @return 返回 threadAlive
 */
public boolean isThreadAlive() {
    return threadAlive;
}
/**
 * @param threadAlive
 *              要设置的 threadAlive
 */
public void setThreadAlive(boolean threadAlive) {
    this.threadAlive = threadAlive;
}
public void active(){
    this.showNotify();
}
public void unActive(){
    this.hideNotify();
}
}}
```

6.6 本章小结

本章主要介绍了与 MIDP 2.0 游戏开发相关的类，比如 Layer、LayerManager、Sprite 和 TiledLayer 等类的使用，读者需要掌握这几个类的基本用法，懂得这几个类的作用，后面章节的游戏开发主要就是围绕这几个类进行的。

7 手机 RPG 游戏设计与实现

本章主要介绍 Java ME 手机射击类游戏的设计与实现。读者需要掌握以下知识点：
- 游戏进度条的实现。
- 游戏主菜单的实现。
- 游戏异常处理。
- 游戏玩家排行榜。
- 主角和怪物的实现。
- 场景切换的实现。
- 游戏 BOSS 的实现。
- 技能魔法的实现。
- 道具系统的实现。

7.1 游戏概述

本章我们将学习一个垂直滚动射击游戏的制作，主要目的是理解一个典型的游戏的编码规范，同时也将前面所学知识包括 MIDP 2.0 的类结合到实际应用中，以及处理移动设备自身限制所带来的挑战。虽然这个游戏不一定是当下最流行的游戏，但是它将揭示开发一个完整游戏从开始到结束的基本过程。

本游戏是类似于"雷电"的竖直滚动的射击游戏，游戏的主要特征如下：
- 基于使用 TiledLayer 类预定义好的地图的三级关卡
- 每级关卡都有一个老怪（BOSS）
- 两个供选择船只（玩家控制）
- 自动存档

- 最高得分（本地或网络同步）
- 各种不同的武器（Wing Man、炸弹 Bombs、Spread、激光 Laser 和防护罩 Shield 等）

7.2 游戏启动画面

启动画面通常在主菜单显示之前，游戏加载时出现。虽然启动画面不是必须的，但是它增加了游戏的吸引力。启动画面一般显示一些图片、公司信息和公司注册商标，可以用 Alert 类简单生成，也可以结合 Timer 类和 Canvas 类定制启动画面。启动画面只需显示很短的时间，一般为 3 秒。在用户化的启动画面中最好有选项可以使用户跳过启动画面立即进入游戏主菜单。

这里直接使用了 Timer 类和 Canves 类，而时间倒计时控制则使用了另外一个名为 CountDown 的类。当然，启动画面本身没有任何意义，需要主类 Midlet 调用才有意义。详细代码如下：

```
import java.util.Timer;

import javax.microedition.lcdui.Canvas;
import javax.microedition.lcdui.Display;
import javax.microedition.lcdui.Displayable;
import javax.microedition.lcdui.Graphics;
import javax.microedition.lcdui.Image;

public final class SplashScreen extends Canvas {
    private Display display;

    private Displayable next;

    private Timer timer;

    private Image image;

    private int dismissTime;

    public SplashScreen(Display display, Displayable next, Image image,
            int dismissTime) {
        timer = new Timer();
        this.display = display;
        this.next = next;
        this.image = image;
```

```java
            this.dismissTime = dismissTime;
            display.setCurrent(this);
        }

        static void access(SplashScreen splashScreen) {
            splashScreen.dismiss();
        }

        private void dismiss() {
            timer.cancel();
            display.setCurrent(next);
        }

        protected void keyPressed(int keyCode) {
            dismiss();
        }

        protected void paint(Graphics g) {
            g.setColor(0x00FFFFFF);
            g.fillRect(0, 0, getWidth(), getHeight());
            g.setColor(0x00000000);
            g.drawImage(image, getWidth() / 2, getHeight() / 2 - 5, 3);
        }

        protected void pointerPressed(int x, int y) {
            dismiss();
        }

        protected void showNotify() {
            if (dismissTime > 0)
                timer.schedule(new CountDown(this), dismissTime);
        }
    }

import java.util.TimerTask;

class CountDown extends TimerTask {
    private final SplashScreen splashScreen;

    CountDown(SplashScreen splashScreen) {
```

```
            this.splashScreen = splashScreen;
        }

        public void run() {
            SplashScreen.access(this.splashScreen);
        }
    }
```

7.3　游戏主菜单的实现

主菜单无论是对游戏还是对应用程序都很重要，在移动设备上更是如此。造就一款成功的游戏，抛开其他各种因素不谈，主菜单的设计尤为重要，其关键是可用性。

主菜单的可用性是指高效、易用、易学、易记。如果一个邮件客户端难以使用或者用户界面不友好，无论定价多低，用户都不会使用。在游戏中，如果用户界面不友好就更糟，即使玩家对游戏感兴趣，由于菜单的原因使游戏难以控制，玩家也不会再玩这个游戏。一个可用的主菜单是非常关键的，所以主菜单系统必须在游戏设计时就进行。而且如果第一次就做得比较好，在以后的游戏开发中可以继续沿用相同的主菜单。

本游戏的主菜单为：
- 开始游戏
- 游戏设置
- 游戏帮助
- 关于我们
- 退出游戏

游戏主菜单界面如图 7-1 所示。

图 7-1　游戏主菜单

游戏主菜单实现源代码如下：

```java
import java.io.IOException;

import javax.microedition.lcdui.Graphics;
import javax.microedition.lcdui.Image;
import javax.microedition.lcdui.game.GameCanvas;
import javax.microedition.lcdui.game.Sprite;
import javax.microedition.midlet.MIDletStateChangeException;

public class Menu extends GameCanvas implements InitData {
    /**
     * 游戏菜单类（提供玩家选择游戏选项）
     */
    MainMIDlet main = null;
    Graphics g = null;
    Image imgMenu = null;
    Image imgMenuItem = null;
    Image imgMenuBg = null;
    Image imgs = null;
    Sprite menuSprite = null;
    int frame = 0;
    int x = 0;
    int y = 0;
    Music music = null;
    int musicFlag=0;
    public Menu(MainMIDlet main, Music music) {
        super(false);
        this.music = music;
        this.main = main;
        g = this.getGraphics();
        this.init();
    }

    //游戏资源初始化
    public void init() {
        music.playBgMusic();
        x = (W - 116) / 2;
        y = H - 60;
        try {
            imgMenuBg = Image.createImage("/black.png");
```

```java
                imgMenuItem = Image.createImage("/menu_1.png");
                imgMenu = Image.createImage("/menu_2.png");
        } catch (Exception e) {
                e.printStackTrace();
        }
        menuSprite = new Sprite(imgMenu, 73, 20);
        menuSprite.setPosition(x + 21, y + 5);
        this.draw();
}

//绘制屏幕内容
public void draw() {
        //清屏
        g.setColor(0, 0, 0);
        g.fillRect(0, 0, W, H);
        //绘制背景图片
        g.drawImage(imgMenuBg, 0, 0, g.LEFT | g.TOP);
        //绘制菜单框
        g.drawImage(imgMenuItem, x, y, g.LEFT | g.TOP);
        //绘制菜单
        menuSprite.setFrame(frame);
        menuSprite.paint(g);
        this.flushGraphics();
}

public void keyPressed(int keyCode) {
        int action = this.getGameAction(keyCode);
        switch (action) {
        case GameCanvas.LEFT:
                music.playMenuMusic();
                frame--;
                if (frame < 0) {
                        frame = 4;
                }
                break;
        case GameCanvas.RIGHT:
                music.playMenuMusic();
                frame++;
                if (frame > 4) {
                        frame = 0;
                }
```

```java
                break;
        case GameCanvas.FIRE:
            music.playMenuMusic();
            switch(frame){
            case 0:
                //开始游戏
                main.display.setCurrent(new MainGame(main,music));
                break;
            case 1:
                //游戏帮助
                main.display.setCurrent(new Help(main,this));
                break;
            case 2:
                //关于游戏
                main.display.setCurrent(new About(main,this));
                break;
            case 3:
                //音乐开关
                musicFlag++;
                if(musicFlag%2==0){
                    music.isPlay=true;
                    music.playBgMusic();
                }else{
                    music.stopBgMusic();
                    music.isPlay=false;
                }
                break;
            case 4:
                //退出游戏
                try {
                        main.destroyApp(true);
                } catch (MIDletStateChangeException e) {
                        e.printStackTrace();
                }
                break;
            }
            break;
        }
        draw();
    }
}
```

按键盘左右键可以选择不同的菜单项。当然，菜单还有其他不同的样式，比如说列表式的菜单，代码的写法类似。

7.4 "关于我们"菜单的实现

当用户选择主菜单中的"关于我们"选项后，游戏跳出一个界面显示有关本游戏的一些信息提示，如图 7-2 所示。主要实现代码如下：

```java
import javax.microedition.lcdui.Canvas;
import javax.microedition.lcdui.Graphics;

public class About extends Canvas implements Runnable, InitData {
    /**
     * 关于我们（游戏开发组相关说明和版权）
     */
    int index=0;
    boolean isAlive=true;
    String msg=
            "=====XXXX 游戏开发小组=====\n"+
            "策划：张三\n"+
            "美工：李四\n"+
            "程序：王二麻子\n"+
            "联系电话：130110110110\n"+
            "QQ：234242342\n"+
            "Email：aaaa@sina.com\n"+
            "地址：湖南长沙星沙湖南大众传媒学院\n";
    char chr[]=null;
    int flag=0;
    int x=10;
    int y=10;
    int width=10;
    int height=20;
    MainMIDlet main=null;
    Menu menu=null;
    public About(MainMIDlet main,Menu menu) {
        this.main=main;
        this.menu=menu;
        chr=msg.toCharArray();
        new Thread(this).start();
```

```
}
public void keyPressed(int keyCode){
    main.display.setCurrent(menu);
}
protected void paint(Graphics g) {
    //输出黑色背景色
    if(flag==0){
        g.setColor(0,0,0);
        g.fillRect(0, 0, W, H);
        flag++;
    }
    //输出信息
    g.setColor(255,255,255);
    g.drawString(chr[index]+"", x, y, g.LEFT|g.TOP);
    x+=width;
    if((chr[index]=='\n')||(x>=W)){
        x=10;
        y+=height;
    }
    index++;
    if(index>=chr.length){
        isAlive=false;
    }
}

public void run() {
    while(isAlive){
        long beginTime = System.currentTimeMillis();
        repaint();
        long endTime = System.currentTimeMillis();
        try {
            Thread.sleep(83 - (endTime - beginTime));
        } catch (InterruptedException e) {
            e.printStackTrace();
        }
    }
}
}
```

图 7-2 "关于我们"菜单

以上字符的出现类似于打字机的效果。

7.5 "游戏帮助"菜单的实现

当玩家选择游戏主菜单中的"游戏帮助"选项后,游戏跳出帮助信息来指导玩家进行游戏,如图 7-3 所示,主要实现代码如下:

```
import java.io.IOException;

import javax.microedition.lcdui.Canvas;
import javax.microedition.lcdui.Graphics;
import javax.microedition.lcdui.Image;

public class Help extends Canvas implements Runnable,InitData{
/**
 * 游戏帮助类
 */
    boolean isAlive=true;
    Image img=null;
    MainMIDlet main=null;
    int size=20;
    int height=15;
    Menu menu=null;
    public Help(MainMIDlet main,Menu menu){
```

```java
        this.main=main;
        this.menu=menu;
        this.init();
        new Thread(this).start();
    }
    public void init(){
        try {
            img=Image.createImage("/help.png");
        } catch (IOException e) {
            //TODO Auto-generated catch block
            e.printStackTrace();
        }
    }
    public void paint(Graphics g){
        //清屏
        g.setColor(255,255,255);
        g.fillRect(0, 0, W, H);
        //绘制帮助图
        g.drawImage(img, 0, 0,g.LEFT|g.TOP);
        //绘制遮盖的百叶窗
        g.setColor(0,255,0);
        for(int i=0;i<size;i++){
            g.fillRect(0,i*15,W,height);
        }
    }
    public void run(){
        while(isAlive){
            long beginTime = System.currentTimeMillis();
            height--;
            if(height<=0){
                height=0;
                isAlive=false;
            }
            repaint();
            long endTime = System.currentTimeMillis();
            try {
                Thread.sleep(83 - (endTime - beginTime));
            } catch (InterruptedException e) {
                e.printStackTrace();
            }
```

```
            }
        }
        public void keyPressed(int keyCode){
            main.display.setCurrent(menu);
        }
    }
```

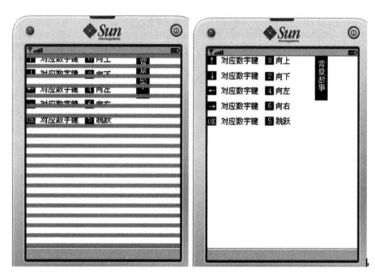

图 7-3 "游戏帮助"菜单

帮助界面的内容首先是出现一个百叶窗的效果，百叶窗慢慢消失后再逐渐显示帮助信息的主体内容。

7.6 "游戏设置"菜单的实现

本游戏暂时只实现了游戏音乐开关的设置，其他功能的写法与此类似，本章不再详述。音乐开关设置的源代码如下：

```
import java.io.IOException;
import java.io.InputStream;

import javax.microedition.media.Manager;
import javax.microedition.media.MediaException;
import javax.microedition.media.Player;

public class Music {
    /**
```

```java
 * 游戏音乐音效类
 */
Player player[] = null;//0=背景音乐 1=菜单音效 2=魔法攻击 3=物理攻击

boolean isPlay = true;

InputStream in = null;

public Music() {
    player = new Player[4];
    this.initSource();
}

public void initSource() {
    try {
        //初始化背景音乐
        in = this.getClass().getResourceAsStream("bg.mid");
        player[0] = Manager.createPlayer(in, "audio/midi");
        player[0].setLoopCount(-1);
        player[0].realize();
        player[0].prefetch();
        //初始化菜单音效
        in = this.getClass().getResourceAsStream("menu.wav");
        player[1] = Manager.createPlayer(in, "audio/x-wav");
        player[1].realize();
        player[1].prefetch();
        //初始化魔法攻击音效
        in = this.getClass().getResourceAsStream("mofa.wav");
        player[2] = Manager.createPlayer(in, "audio/x-wav");
        player[2].realize();
        player[2].prefetch();
        //初始化物理攻击音效
        in = this.getClass().getResourceAsStream("gongji.wav");
        player[3] = Manager.createPlayer(in, "audio/x-wav");
        player[3].realize();
        player[3].prefetch();
    } catch (IOException e) {
        e.printStackTrace();
    } catch (MediaException e) {
        e.printStackTrace();
    }
}
```

```java
    }

    //播放背景音乐
    public void playBgMusic() {
        if (isPlay) {
            if (player[0].getState() != Player.STARTED) {
                try {
                    //player[0].start();
                } catch (Exception e) {
                    e.printStackTrace();
                }
            }
        }
    }

    //关闭背景音乐
    public void stopBgMusic() {
        if (isPlay) {
            isPlay = false;
            if (player[0].getState() == Player.STARTED) {
                try {
                    player[0].stop();
                } catch (MediaException e) {
                    e.printStackTrace();
                }
            }
        }
    }

    //==========菜单音效====================
    //播放菜单音乐
    public void playMenuMusic() {
        if (isPlay) {
            if (player[1].getState() != Player.STARTED) {
                try {
                    player[1].start();
                } catch (MediaException e) {
                    e.printStackTrace();
                }
            }
```

```java
    }

    //关闭菜单音乐
    public void stopMenuMusic() {
        if (isPlay) {
            isPlay = false;
            if (player[1].getState() == Player.STARTED) {
                try {
                    player[1].stop();
                } catch (MediaException e) {
                    e.printStackTrace();
                }
            }
        }
    }

    //=========魔法攻击音效====================
    //播放魔法攻击音乐
    public void playMofaMusic() {
        if (isPlay) {
            if (player[2].getState() != Player.STARTED) {
                try {
                    player[2].start();
                } catch (MediaException e) {
                    e.printStackTrace();
                }
            }
        }
    }

    //关闭魔法攻击音乐
    public void stopMofaMusic() {
        if (isPlay) {
            isPlay = false;
            if (player[2].getState() == Player.STARTED) {
                try {
                    player[2].stop();
                } catch (MediaException e) {
                    e.printStackTrace();
                }
            }
```

```java
            }
        }

        //=========物理攻击音效==================
        //播放物理攻击音乐
        public void playGongjiMusic() {
            if (isPlay) {
                if (player[3].getState() != Player.STARTED) {
                    try {
                        player[3].start();
                    } catch (MediaException e) {
                        e.printStackTrace();
                    }
                }
            }
        }

        //关闭物理攻击音乐
        public void stopGongjiMusic() {
            if (isPlay) {
                isPlay = false;
                if (player[3].getState() == Player.STARTED) {
                    try {
                        player[3].stop();
                    } catch (MediaException e) {
                        e.printStackTrace();
                    }
                }
            }
        }
}
```

7.7 怪物敌人功能的实现

玩家在游戏的过程中要不断地杀怪物来获取经验以提升等级，那么怪物怎么实现呢？

其实这可以用前面学过的 Sprite 类来实现，但是有点不同的就是我们要用自定义的 Sprite 类来呈现怪物的样式和动作状态。以下是怪物的实现源代码：

```java
import javax.microedition.lcdui.Image;
import javax.microedition.lcdui.game.Sprite;
/**
```

```java
 * 本类实现怪物
 * @author yc
 *
 */
public class EmyClass extends Sprite implements Runnable, InitData {
    boolean isAlive = true;

    int move[] = { 0, 1, 2, 3 };

    int fire[] = { 4, 5, 6, 7 };

    int action = 0;//0=移动状态  1=攻击和被攻击状态
    int nowAction=0;//当前状态
    int flag = 0;
    int speed=emy_speed_1;
    public EmyClass(Image img, int w, int h, int x, int y) {
        super(img, w, h);
        this.setPosition(x, y);
        this.setFrameSequence(move);
        this.defineReferencePixel(w / 2, h / 2);
        this.setTransform(Sprite.TRANS_MIRROR);
        new Thread(this).start();
    }

    public void run() {
        while (isAlive) {
            long beginTime = System.currentTimeMillis();
            this.move(speed, 0);
            flag++;
            if(flag>=20){
                if(speed<0){
                    this.setTransform(Sprite.TRANS_MIRROR);
                }else{
                    this.setTransform(Sprite.TRANS_NONE);
                }
                flag=0;
                speed*=-1;
            }
            this.nextFrame();
            long endTime = System.currentTimeMillis();
```

```
            try {
                    Thread.sleep(83 - (endTime - beginTime));
            } catch (InterruptedException e) {
                    e.printStackTrace();
            }
        }
    }

}
```

7.8 怪物 BOSS 功能的实现

怪物 BOSS 和普通小怪的实现效果差不多，唯一不同的地方就是 BOSS 拥有更多的 AI，所以程序也主要是实现 BOSS 的 AI 动作。源代码如下：

```
import javax.microedition.lcdui.Image;
import javax.microedition.lcdui.game.Sprite;
/**
 * 此类为怪物 BOSS 类
 * @author Administrator
 *
 */
public class Boss extends Sprite implements Runnable{
    boolean isAlive=true;
    int x=180;
    int y=210;
    int moveFrame[]={0,1,2};
    int fireFrame[]={3,4,5,6,7,8};
    Sprite npcSprite=null;
    int speed=3;
    int dir=0;//0=left   1=right
    int action=0;//0=移动追击   1=攻击
    public Boss(Image img,Sprite npcSprite){
        super(img,60,50);
        this.npcSprite=npcSprite;
        this.setPosition(180,210);
        this.defineReferencePixel(30, 25);
        this.setFrameSequence(moveFrame);
        new Thread(this).start();
    }
    public void ai(){
```

```java
        //获取主角人物坐标
        int npcX=npcSprite.getX();
        int npcY=npcSprite.getY();
        //判定 boss 追踪的方向
        if(npcX>this.getX()+10){
            //往右边追
            if(action!=0){
                this.setFrameSequence(moveFrame);
            }
            action=0;
            this.setTransform(Sprite.TRANS_MIRROR);
            this.move(speed, 0);
        }else if(npcX<this.getX()-10){
            //往左边追
            if(action!=0){
                this.setFrameSequence(moveFrame);
            }
            action=0;
            this.setTransform(Sprite.TRANS_NONE);
            this.move(-speed, 0);
        }
        if(this.collidesWith(npcSprite, false)){
            if(action!=1){
                this.setFrameSequence(fireFrame);
            }
            action=1;
        }
    }
    public void run(){
        while(isAlive){
            this.ai();
            //this.nextFrame();
            try {
                Thread.sleep(83);
            } catch (InterruptedException e) {
                e.printStackTrace();
            }
        }
    }
}
```

7.9 人物魔法技能功能的实现

在手机游戏中，人物魔法技能的实现也可以用自定义 Sprite 类，主要实现代码如下：

```java
import javax.microedition.lcdui.Image;
import javax.microedition.lcdui.game.Sprite;
/**
 * 魔法技能类
 * @author Administrator
 *
 */
public class Magic extends Sprite implements Runnable,InitData {
    boolean isAlive=true;
    int frame[]={3,3,4,4,4,5,5,5,6,6,6,6,6};
    int dir;
    int speed=8;
    public Magic(Image img,int w,int h,int x,int y,int dir){
        super(img,w,h);
        this.dir=dir;
        this.setFrameSequence(frame);
        this.setPosition(x, y);
        this.defineReferencePixel(8,8);
        new Thread(this).start();
    }
    public void run(){
        while(isAlive){
            switch(dir){
            case 1:
                //右飞
                this.move(speed,0);
                break;
            case -1:
                //左飞
                this.move(-speed,0);
                break;
            }
            //魔法没打中，超出边界自动消亡
            if((this.getX()<=-30)||(this.getX()>=W+30)){
                isAlive=false;
            }
```

```
                this.nextFrame();
                try {
                    Thread.sleep(83);
                } catch (InterruptedException e) {
                    e.printStackTrace();
                }
            }
        }
    }
}
```

7.10 游戏碰撞检测功能的实现

在手机 RPG 游戏（Role-playing Gome）中，碰撞检测是必须要实现的一个主要功能。本游戏中主要的碰撞检测类型包括：

- 人物角色与怪物碰撞
- 人物角色与地图碰撞
- 魔法技能与怪物碰撞
- 魔法技能与地图碰撞

碰撞检测主要实现代码如下：

```
//游戏碰撞检测
public boolean collids(){
    boolean flag=false;
    //人物角色的边界碰撞
    if((npcSprite.getX()>(W-30))||(npcSprite.getX()<0)||(npcSprite.getY()<y_1)||(npcSprite.getY()>H-60)){
        flag=true;
    }
    return flag;
}
//判定 boss 攻击人物掉血
public void bossFireNpc(){
    if(boss!=null){
        System.out.println(action+"    "+boss.getFrame());
        if((boss.action==1)&&(boss.getFrame()>=5)){
            nowHP-=5;
        }
        boss.nextFrame();
    }
}
//人物与道具碰撞
```

```java
public void npcCollidsDaojv(){
    if(daojv!=null){
        if(npcSprite.collidesWith(daojv, false)){
            if(!daojv.collids){
                switch(id){
                case 0:
                    //加血 20
                    nowHP+=20;
                    if(nowHP>=HP){
                        nowHP=HP;
                    }
                    break;
                case 1:
                    //加魔法 10
                    nowMP+=10;
                    if(nowMP>=MP){
                        nowMP=MP;
                    }
                    break;
                }
            }
            daojv.collids=true;
        }
        if(!daojv.isAlive){
            daojv=null;
        }
    }
}
//魔法、人物与怪物碰撞检测
public void fireCollids(){
//魔法技能与怪物碰撞检测
    if((emy!=null)&&(magic!=null)){
        if(magic.collidesWith(emy, false)){
            id=rd.nextInt(idSize);
            //获取当前道具掉落的几率表
            String diaoluo="";
            switch(id){
            case 0:
                diaoluo=diao_hp_1;
                break;
```

```
            case 1:
                diaoluo=diao_mp_1;
                break;
        }
        //根据掉落几率字符串分析出掉落的参数
        int pos=diaoluo.indexOf("-");
        int fenzi=Integer.parseInt(diaoluo.substring(0,pos));
        int fenmu=Integer.parseInt(diaoluo.substring(pos+1,diaoluo.length()));
        if(rd.nextInt(fenmu)<fenzi){
            daojv=new DaoJv(imgDaoJv,15,15,emy.getX()+5,emy.getY()+15,id);
        }
        //清除敌人和魔法
        emy=null;
        magic=null;
        emySize--;
        if(emySize<=0){
            boss=new Boss(imgBoss,npcSprite);
        }
    }
}
if(emy!=null){
    if(npcSprite.collidesWith(emy, false)){
        nowHP-=5;
        if(nowHP<=0){
            nowHP=0;
        }
    }
}
}
```

7.11 游戏按键检测功能的实现

Java ME 手机游戏中玩家与游戏互动主要靠键盘响应玩家的按键来实现,当然,触摸屏事件也是属于这一类型的事件响应。本游戏只以实现键盘按键做实例说明,源代码如下:

```
//游戏按键控制方法
public void input(){
    int state=this.getKeyStates();
    if((state & GameCanvas.LEFT_PRESSED)!=0){
        //按左方向键
        if(action!=0){
```

```java
            action=0;
            npcSprite.setFrameSequence(npcMove);
        }
        dir=-1;
        //踩雷,随机刷怪
        this.makeEmy();
        npcSprite.setTransform(Sprite.TRANS_NONE);
        if(mapBgSprite[0].getX()>=0){
            npcSprite.move(-speed_1, 0);
        }else if(emy==null){
                for(int i=0;i<mapBgSprite.length;i++){
                    mapBgSprite[i].move(speed_1, 0);
                }
        }
        npcSprite.nextFrame();
        if(this.collids()){
            npcSprite.move(speed_1, 0);
        }
    }else if((state & GameCanvas.RIGHT_PRESSED)!=0){
        //向右方向键
        if(action!=0){
            action=0;
            npcSprite.setFrameSequence(npcMove);
        }
        dir=1;
        //踩雷,随机刷怪
        this.makeEmy();
        npcSprite.setTransform(Sprite.TRANS_MIRROR);
        if(npcSprite.getX()<=(W-90)){
            npcSprite.move(speed_1, 0);
        }else   if(emy==null){
            for(int i=0;i<mapBgSprite.length;i++){
                mapBgSprite[i].move(-speed_1, 0);
            }
        }
        npcSprite.nextFrame();
        if(mapBgSprite[0].getX()<=-200){
            for(int i=0;i<mapBgSprite.length;i++){
                mapBgSprite[i].setPosition(i*200, 0);
            }
        }
```

```
        }
        if(this.collids()){
            npcSprite.move(-speed_1, 0);
        }
    }else if((state & GameCanvas.UP_PRESSED)!=0){
        //按上方向键
        if(action!=0){
            action=0;
            npcSprite.setFrameSequence(npcMove);
        }
        npcSprite.move(0, -speed_1);
        npcSprite.nextFrame();
        if(this.collids()){
            npcSprite.move(0, speed_1);
        }
    }else if((state & GameCanvas.DOWN_PRESSED)!=0){
        //按下方向键
        if(action!=0){
            action=0;
            npcSprite.setFrameSequence(npcMove);
        }
        npcSprite.move(0, speed_1);
        npcSprite.nextFrame();
        if(this.collids()){
            npcSprite.move(0, -speed_1);
        }
    }else if((state & GameCanvas.FIRE_PRESSED)!=0){
        //按攻击键
        if(action!=1){
            action=1;
            npcSprite.setFrameSequence(npcFire);
        }
        if(skillCD<=0){
            skillCD=40;
            nowMP-=10;
            if(nowMP>=10){
                music.playMofaMusic();
                magic=new Magic(imgMagic,16,16,npcSprite.getX()+30,npcSprite.getY()+30,dir);
            }
        }
```

```
        }
        if(action==1){
            npcSprite.nextFrame();
        }
    }
```

7.12 游戏主要逻辑循环功能的实现

游戏主要逻辑循环是整个游戏运行的核心，主要实现代码如下：

```java
import java.util.Random;

import javax.microedition.lcdui.Graphics;
import javax.microedition.lcdui.Image;
import javax.microedition.lcdui.game.GameCanvas;
import javax.microedition.lcdui.game.LayerManager;
import javax.microedition.lcdui.game.Sprite;

public class MainGame extends GameCanvas implements Runnable,InitData {
    /**
     * 游戏主循环
     */
    MainMIDlet main=null;
    Music music=null;
    boolean isAlive=true;
    Graphics g=null;
    LayerManager layer=null;
    //人物角色资源和参数
    Image imgNPC=null;
    Sprite npcSprite=null;
    int npcMove[]={0,1,2,3,4};
    int npcFire[]={4,5,6,7};
    //游戏背景图资源和参数
    Image imgMapBg=null;
    Sprite mapBgSprite[]=null;
    //游戏怪物资源
    Image imgEmy=null;
    int emySize=1;
    Random rd=null;
    EmyClass emy=null;
    //人物角色状态信息资源和参数
```

```java
Image imgInfo=null;
Sprite hpSprite=null;
Sprite mpSprite=null;
int infoX=0;
int infoY=5;
int nowHP=HP;
int nowMP=MP;
int action=0;//0=移动状态    1=攻击状态
int skillCD=0;//技能 CD
//魔法技能
Image    imgMagic=null;
Magic magic=null;
int dir=1;//人物方向  dir=1    右    dir=-1  左
//道具图片资源和参数
Image imgDaoJv=null;
int idSize=2;//物品 ID 总数
int id=0;//当前掉落的物品 ID
DaoJv daojv=null;
//BOSS 图片资源和相关参数
Image imgBoss=null;
Boss boss=null;
public MainGame(MainMIDlet main,Music music){
    super(true);
    this.main=main;
    this.music=music;
    this.init();
    new Thread(this).start();
}
//游戏资源初始化
public void init(){
    layer=new LayerManager();
    g=this.getGraphics();
    try {
        //初始化人物角色资源
        imgNPC=Image.createImage("/npc.png");
        npcSprite=new Sprite(imgNPC,60,60);
        npcSprite.setPosition(x_1, y_1);
        npcSprite.setFrameSequence(npcMove);
        npcSprite.defineReferencePixel(30, 30);
        layer.append(npcSprite);
        imgInfo=Image.createImage("/info.png");
```

```java
            hpSprite=new Sprite(imgInfo,150,15);;
            hpSprite.setPosition(infoX,infoY);
            hpSprite.setFrame(0);
            mpSprite=new Sprite(imgInfo,150,15);;
            mpSprite.setPosition(infoX,infoY+15);
            mpSprite.setFrame(1);
            //初始化游戏背景地图资源
            imgMapBg=Image.createImage("/mapbg.png");
            mapBgSprite=new Sprite[3];
            for(int i=0;i<mapBgSprite.length;i++){
                mapBgSprite[i]=new Sprite(imgMapBg,200,320);
                mapBgSprite[i].setFrame(0);
                mapBgSprite[i].setPosition(i*200, 0);
                layer.append(mapBgSprite[i]);
            }
            //初始化怪物资源
            rd=new Random();
            imgEmy=Image.createImage(img_file_name_1);
            //初始化魔法技能资源
            imgMagic=Image.createImage("/magic.png");
            //初始化物品道具资源
            imgDaoJv=Image.createImage("/daojv.png");
            //初始化 boss 图片资源
            imgBoss=Image.createImage("/boss.png");
        } catch (Exception e) {
            e.printStackTrace();
        }
    }
    //随机刷怪
    public void makeEmy(){
        int temp=rd.nextInt(100);
        if((temp<2)&&(emy==null)&&(emySize>0)){
            emy=new EmyClass(imgEmy,emy_w_1,emy_h_1,rd.nextInt(W-emy_w_1),rd.nextInt(20)+200);
        }
    }
    //游戏绘制方法
    public void draw(){
        //清屏
        g.setColor(255,255,255);
        g.fillRect(0, 0, W, H);
        //绘制游戏资源
```

```java
            layer.paint(g, 0, 0);
            //绘制人物角色信息
            g.setColor(0,0,255);
            g.fillRect(infoX+35,infoY+3,(nowMP*100/MP),10);
            g.setColor(255,0,0);
            g.fillRect(infoX+35,infoY+17,(nowHP*100)/HP,10);
            hpSprite.paint(g);
            mpSprite.paint(g);
            //绘制怪物
            if(emy!=null){
                  emy.paint(g);
            }
            //绘制技能 CD 图标
            g.setColor(0,0,255);
            g.fillRect((W-skillCD)/2, 250,skillCD,5);
            if(skillCD>=0){
                  skillCD-=2;
            }
            //绘制魔法技能
            if(magic!=null){
                  magic.paint(g);
            }
            //绘制掉落的道具
            if(daojv!=null){
                  daojv.paint(g);
            }
            //绘制 boss
            if(boss!=null){
                  boss.paint(g);
            }
      }
      //游戏按键控制方法
      public void input(){
            int state=this.getKeyStates();
            if((state & GameCanvas.LEFT_PRESSED)!=0){
                  //按左方向键
                  if(action!=0){
                        action=0;
                        npcSprite.setFrameSequence(npcMove);
                  }
                  dir=-1;
```

```java
            //踩雷,随机刷怪
            this.makeEmy();
            npcSprite.setTransform(Sprite.TRANS_NONE);
            if(mapBgSprite[0].getX()>=0){
                npcSprite.move(-speed_1, 0);
            }else if(emy==null){
                    for(int i=0;i<mapBgSprite.length;i++){
                        mapBgSprite[i].move(speed_1, 0);
                    }
            }
            npcSprite.nextFrame();
            if(this.collids()){
                npcSprite.move(speed_1, 0);
            }
        }else if((state & GameCanvas.RIGHT_PRESSED)!=0){
            //向右方向键
            if(action!=0){
                action=0;
                npcSprite.setFrameSequence(npcMove);
            }
            dir=1;
            //踩雷,随机刷怪
            this.makeEmy();
            npcSprite.setTransform(Sprite.TRANS_MIRROR);
            if(npcSprite.getX()<=(W-90)){
                npcSprite.move(speed_1, 0);
            }else    if(emy==null){
                for(int i=0;i<mapBgSprite.length;i++){
                    mapBgSprite[i].move(-speed_1, 0);
                }
            }
            npcSprite.nextFrame();
            if(mapBgSprite[0].getX()<=-200){
                for(int i=0;i<mapBgSprite.length;i++){
                    mapBgSprite[i].setPosition(i*200, 0);
                }
            }
            if(this.collids()){
                npcSprite.move(-speed_1, 0);
            }
        }else if((state & GameCanvas.UP_PRESSED)!=0){
```

```
        //按上方向键
        if(action!=0){
            action=0;
            npcSprite.setFrameSequence(npcMove);
        }
        npcSprite.move(0, -speed_1);
        npcSprite.nextFrame();
        if(this.collids()){
            npcSprite.move(0, speed_1);
        }
    }else if((state & GameCanvas.DOWN_PRESSED)!=0){
        //按下方向键
        if(action!=0){
            action=0;
            npcSprite.setFrameSequence(npcMove);
        }
        npcSprite.move(0, speed_1);
        npcSprite.nextFrame();
        if(this.collids()){
            npcSprite.move(0, -speed_1);
        }
    }else if((state & GameCanvas.FIRE_PRESSED)!=0){
        //按攻击键
        if(action!=1){
            action=1;
            npcSprite.setFrameSequence(npcFire);
        }
        if(skillCD<=0){
            skillCD=40;
            nowMP-=10;
            if(nowMP>=10){
                music.playMofaMusic();
                magic=new Magic(imgMagic,16,16,npcSprite.getX()+30,npcSprite.getY()+30,dir);
            }
        }
    }
    if(action==1){
        npcSprite.nextFrame();
    }
}
//游戏碰撞检测
```

```java
public boolean collids(){
    boolean flag=false;
    //人物角色的边界碰撞
    if((npcSprite.getX()>(W-30))||(npcSprite.getX()<0)||(npcSprite.getY()<y_1)||(npcSprite.getY()>H-60)){
        flag=true;
    }
    return flag;
}
//判定boss攻击人物掉血
public void bossFireNpc(){
    if(boss!=null){
        System.out.println(action+"    "+boss.getFrame());
        if((boss.action==1)&&(boss.getFrame()>=5)){
            nowHP-=5;
        }
        boss.nextFrame();
    }
}
//人物与道具碰撞
public void npcCollidsDaojv(){
    if(daojv!=null){
        if(npcSprite.collidesWith(daojv, false)){
            if(!daojv.collids){
                switch(id){
                case 0:
                    //加血20
                    nowHP+=20;
                    if(nowHP>=HP){
                        nowHP=HP;
                    }
                    break;
                case 1:
                    //加魔法10
                    nowMP+=10;
                    if(nowMP>=MP){
                        nowMP=MP;
                    }
                    break;
                }
            }
            daojv.collids=true;
```

```java
            }
            if(!daojv.isAlive){
                daojv=null;
            }
        }
    }
    //魔法、人物与怪物碰撞检测
    public void fireCollids(){
        //魔法技能与怪物碰撞检测
        if((emy!=null)&&(magic!=null)){
            if(magic.collidesWith(emy, false)){
                id=rd.nextInt(idSize);
                //获取当前道具掉落的几率表
                String diaoluo="";
                switch(id){
                case 0:
                    diaoluo=diao_hp_1;
                    break;
                case 1:
                    diaoluo=diao_mp_1;
                    break;
                }
                //根据掉落几率字符串分析出掉落的参数
                int pos=diaoluo.indexOf("-");
                int fenzi=Integer.parseInt(diaoluo.substring(0,pos));
                int fenmu=Integer.parseInt(diaoluo.substring(pos+1,diaoluo.length()));
                if(rd.nextInt(fenmu)<fenzi){
                    daojv=new DaoJv(imgDaoJv,15,15,emy.getX()+5,emy.getY()+15,id);
                }
                //清除敌人和魔法
                emy=null;
                magic=null;
                emySize--;
                if(emySize<=0){
                    boss=new Boss(imgBoss,npcSprite);
                }
            }
        }
        if(emy!=null){
            if(npcSprite.collidesWith(emy, false)){
                nowHP-=5;
```

```java
                    if(nowHP<=0){
                        nowHP=0;
                    }
                }
            }
        }
        public void run(){
            while(isAlive){
                long beginTime=System.currentTimeMillis();
                //按键检测
                this.input();
                //魔法技能碰撞检测
                this.fireCollids();
                //人物与道具碰撞
                this.npcCollidsDaojv();
                //boss 攻击 npc
                this.bossFireNpc();
                //绘制游戏资源
                this.draw();
                this.flushGraphics();
                long endTime=System.currentTimeMillis();
                try {
                    Thread.sleep(83-(endTime-beginTime));
                } catch (InterruptedException e) {
                    e.printStackTrace();
                }
            }
        }
    }
```

7.13 其他功能的实现

7.13.1 游戏加载进度条类

```java
import java.util.Random;
import javax.microedition.lcdui.Canvas;
import javax.microedition.lcdui.Graphics;
public class Loading extends Canvas implements Runnable, InitData {
    /**
```

* 游戏进度条（读取游戏资源）
 */
```java
MainMIDlet main = null;

boolean isAlive = true;//控制线程运行

int flag = 0;//控制进度条播放的长度(0~200)

Random rd = null;

Music music = null;

public Loadding(MainMIDlet main) {
    this.main = main;
    rd = new Random();
    new Thread(this).start();
    music = new Music();
}

public void paint(Graphics g) {
    //清屏
    g.setColor(0, 0, 0);
    g.fillRect(0, 0, W, H);
    //绘制一行文字显示 "游戏加载中 x%"
    g.setColor(0, 255, 0);
    g.drawString("游戏资源加载中         %", (W - 105) / 2, (H - 60) / 2, g.LEFT
            | g.TOP);
    g.drawString(Integer.toString(flag / 2), W / 2 + 26, (H - 60) / 2,
            g.LEFT | g.TOP);
    //绘制一个矩形框
    g.setColor(0, 255, 0);
    g.drawRect((W - 200) / 2, (H - 20) / 2, 200, 20);
    //绘制一个填充矩形
    g.fillRect((W - 200) / 2, (H - 20) / 2, flag, 20);
}

public void run() {
    while (isAlive) {
        long beginTime = System.currentTimeMillis();
        flag += (rd.nextInt(90));
```

```java
                    if (flag >= 200) {
                            flag = 200;
                            isAlive = false;
                            //切换到游戏菜单类
                            main.display.setCurrent(new Menu(main, music));
                    }
                    repaint();
                    long endTime = System.currentTimeMillis();
                    try {
                            Thread.sleep(83 - (endTime - beginTime));
                    } catch (InterruptedException e) {
                            e.printStackTrace();
                    }
            }
        }
    }
```

7.13.2 游戏道具类

```java
import javax.microedition.lcdui.Image;
import javax.microedition.lcdui.game.Sprite;

public class DaoJv extends Sprite implements Runnable {
    boolean isAlive=true;
    int frame[]={0,1,2,3};
    boolean collids=false;
    public DaoJv(Image img,int w,int h,int x,int y,int id){
        super(img,w,h);
        this.setPosition(x, y);
        for(int i=0;i<frame.length;i++){
            frame[i]=frame[i]+frame.length*id;
        }
        this.setFrameSequence(frame);
        new Thread(this).start();
    }
    public void run(){
        while(isAlive){
            this.nextFrame();
            if(collids){
                this.move(-2, -20);
```

```
                    if((this.getX()<=0)||(this.getY()<=0)){
                        isAlive=false;
                    }
                }
                try {
                    Thread.sleep(83);
                } catch (InterruptedException e) {
                    e.printStackTrace();
                }
            }
        }
    }
```

7.13.3 游戏公共参数资源配置的实现

```
public interface InitData {
    //人物属性值
    int HP=100;
    int MP=50;
    //游戏画布尺寸
    int W=240;
    int H=291;
    //小怪的血量值
    int SMAL_EMY_HP=100;
    //BOSS 的血量值
    int BOSS_EMY_HP=2000;
    //第一关人物角色参数
    int x_1=20;
    int y_1=190;
    int speed_1=5;

    //第一关怪物参数
    String img_file_name_1="/molong.png";
    int emy_w_1=40;
    int emy_h_1=40;
    int emy_speed_1=2;
    //第二关怪物参数
    String img_file_name_2="/xiezi.png";
    int emy_w_2=60;
    int emy_h_2=60;
```

```
        int emy_speed_2=3;
        //第一关小怪道具掉落列表
        String diao_hp_1="1-2";
        String diao_mp_1="123-500";
}
```

7.14 游戏实现效果图

游戏实现效果图如图 7-4 至图 7-6 所示。

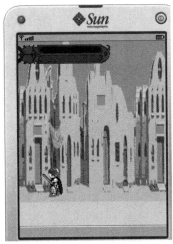

图 7-4　游戏主界面

图 7-5　游戏刷怪

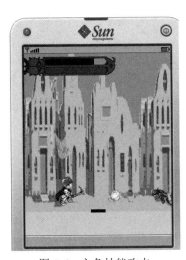

图 7-6　主角技能攻击

7.15 本章小结

本章主要是通过一个手机 RPG 游戏设计与实现的案例来教会读者怎样把前面学到的理论知识运用到实践中开，发一款能实际运行的游戏。读者学完本章后应该对手机 RPG 游戏主循环框架的实现、游戏怪物和魔法技能的实现等方面有所了解。

8 网络编程

本章主要介绍 Java ME 手机网络编程。读者需要掌握以下知识点：
- 连接接口
- 输入/输出流连接接口
- 流连接接口
- 内容连接接口
- HTTP 连接接口
- TCP/IP 协议
- UDP 协议

8.1 移动网络编程概述

本节重点讲解基于 CLDC 的通用连接框架，介绍 javax.microedition.io 包中相关的类和接口，在 Java ME 平台下网络连接的层次结构和联系。

8.1.1 CLDC 通用连接框架

通用连接框架（Generic Connection Framework，GCF）是 Java ME 进行网络通信的基础，它位于 javax.microedition.io 包中，包括一个类、一个异常和七个接口。

通用连接框架中类和接口的层次结构和继承关系如图 8-1 所示。

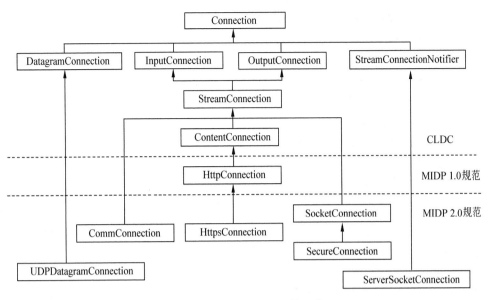

图 8-1　通用连接框架结构层次

Connection 接口是所有接口的基类，它产生了用于数据报连接的 DatagramConnection 连接接口、用于数据流连接的 InputConnection、OutputConnection 和 StreamConnectionNotifier 接口。其中 StreamConnection 接口继承自数据流连接的输入/输出接口，并派生了用于数据内容解析的 ContentConnection 接口。

GCF 是一个容易扩展的框架结构，用户可以根据自身的需要在现有的类和接口之上建立自己的类或者实现自己的接口。

8.1.2　CLDC 通用连接类

下面我们对通用连接框架各种类进行介绍：

1. Connector 类

Connector 类主要用于创建连接接口的对象，该类提供了打开连接和创建输入/输出流的方法。

Public static Connection open(String name);
Public static Connection open(String name, int mode);
Public static Connection open(String name, int mode, boolean timeout);

这三个重载的方法是建立并打开一个连接，参数 name 为 URI（统一资源标记符），用于指定协议和地址，它定义了各种连接的统一格式：

{scheme}: [{target}] [{params}]

这里 scheme 是进行连接的协议，如 HTTP、FTP 等，协议是确定连接具体类型的唯一参数；target 是连接的目标地址，如表示主机的名称或 IP 地址，params 是连接的参数列表，如 user 用户名、password 口令、port 连接端口等。

参数 mode 为访问模式，在 Connector 类中定义了三种取值：

Connector.READ 表示只读模式。

Connector.WRITE 表示只写模式。

Connector.READ_WRITE 表示可读可写模式，默认值为可读可写。

参数 timeout 用于指明调用者是否希望处理超时问题，如果取值为 true，表示希望在连接超时之后触发一个超时异常。

框架支持多种连接类型，通用连接框架中的连接接口都是通过 open()方法建立并通过统一资源统标识符（URI）来确定连接的类型。Open()方法返回一个 Connection 对象，在应用程序中我们总是将返回对象强制转换为特定连接的子类型，如 HttpConnection、SocketConnection 等。

本章主要讲述的三种常用连接类型格式如下：

创建 socket 连接，例如：

SocketConnection c = (SocketConnection)Connector.open("socket://host:1234");

创建数据报连接，例如：

UDPDatagramConnection c =(UDPDatagramConnection) Connector.open ("datagram://host");

创建 http 连接，例如：

HttpConnection c = (HttpConnection)Connector.open("http://mysite.com:8080/index.htm");

由此可见创建各种连接的方法都很类似，只是参数不同。关闭连接的方法是调用连接的 close()方法。

在很多情况下打开连接主要是为了访问特定的输入/输出流,因此 Connector 还提供了建立四种流连接的方法：

（1）Public static DataInputStream openDataInputStream(String name)

建立一个数据输入流，可以进行原始数据类型的输入，参数 name 为 URI（统一资源标记符），用于指定协议和地址。

（2）Public static DataOutputStream openDataOutputStream(String name)

建立一个数据输出流，可以进行原始数据类型的输出，参数 name 为 URI（统一资源标记符），用于指定协议和地址。

（3）Public static IntputStream openInputStram(String name)

建立一个连接输入流，可以进行字节数据类型的输入，参数 name 为 URI（统一资源标记符），用于指定协议和地址。

（4）Public static OutputStream openOutputStram(String name)

建立一个连接输出流，可以进行字节数据类型的输出，参数 name 为 URI（统一资源标记符），用于指定协议和地址。

InputStream 类和 OutputStream 类是所有输入输出类的基类，从这两个类可以派生出特定数据类型的输入输出流。如：ByteArrayOutputStream、ByteArrayInputStream 和 DataInputStream 等。DataInputStream 类和 DataOutputStream 类则可以输入输出 Java 的基本数据类型。

2. Connection 接口

Connection 接口是最基本的连接类型，是其他连接接口的基类，其中定义了 close()方法，用于关闭连接。其语法格式如下：

Public void close();

关闭连接后，如果再进行输入/输出操作则会抛出 IOException 异常。

3. InputConnection 接口和 OutputConnection 接口

InputConnection 接口定义了输入流连接所需的各种方法，在 Connection 的基础上增加了 openInputStream()方法和 openDataInputStream()方法，前者用于打开输入流连接，后者用于打开数据输入流连接。

OutputConnection 接口定义了输出流连接所需的各种方法，增加了 openOutputStream()方法和 openDataOutputStream()方法。

4. StreamConnection 接口

StreamConnection 接口从 InputConnection 和 OutputConnection 派生而来。因此它可以继承前面说明的 openInputStream()方法、openDataInputStream()方法、openOutputStream()方法和 openDataOutputStream()方法，为实现双向通信提供基础。

5. ContentConnection 接口

ContentConnection 接口继承自 StreamConnection 接口，用于获取连接内容的编码、长度和类型等信息。它增加了与连接内容相关的方法：

（1）public String getEncoding()

用于获得连接资源内容编码类型，如果是通过 HTTP 进行连接的，则返回首部字段 content-encoding 的值。

（2）public long getLength()

用于获得内容长度，如果是通过 HTTP 进行连接的，则返回首部字段 content-length 的值。

（3）public String getType()

用于获得内容的资源类型，如果是通过 HTTP 进行连接的，则返回首部字段 content-type 的值。

6. StreamConnectionNotifier 接口

该接口多用于服务器端应用程序，目的是在 Connection 接口的基础上扩展一个 acceptAndOpen()方法，该方法等待客户端的连接。语法格式如下：

Public StreamConnection acceptAndOpen()

返回 StreamConnection 流连接的一个实例，可用于 ServerSocket 通信。

8.2 HTTP 编程

从 MIDP 1.0 规范开始，就要求所有的移动通信设备必须支持 HTTP 协议，因此使用这种协议编程能保证较好的可移植性。所以网络编程首先讲 HTTP 编程，本节用到的 API 主要是：

Javax.microedition.io.HTTPConnection

HTTP 协议是基于请求/响应模式的。该协议中客户机与服务器的信息交换过程如图 8-2 所示，主要分为四个步骤：建立连接、发送请求信息、服务器发送响应信息和关闭连接。

图 8-2 使用 HTTP 协议客户与服务器交换信息的过程

手机的 MIDP 程序与服务器连接，首先需要一个 Web 服务器。本书选用 Tomcat 7.0.2 版本的服务器，安装和设置过程与一般 JSP 系统安装设置过程相同，不再赘述。

8.2.1 MIDLet 连接到 HTTP 服务器上

进行 HTTP 编程首先是将 MIDLet 连接到 HTTP 服务器上，对于 HTTP 连接可使用前面讲过的 Connector 类的静态方法：

public static Connection open(String name) throws IOException

对于其参数 name，为打开 HTTP 连接字符串，格式为：http://IP 地址:端口/资源路径，如：http://127.0.0.1:8080/index.jsp，该函数的返回值需要应用强制类型转换为 HttpConnection 类型。下面的代码将返回一个 MIDlet 与 HTTP 连接的对象：

HttpConnection hc= (HttpConnection)Connctor.open("http://localhost:8066/login.jsp")

建立起来的 HTTP 连接共有三种状态：

- Setup：该状态下可以设置请求参数。
- Connected：在该状态下，请求已经发送出去，正在等待响应。
- Closed：这是 HTTP 连接的最终状态，即连接被关闭。

HTTP 的请求参数必须在请求被发送之前设置完毕。在 HttpConnection 中定义的设置请求的方法有两个：

第一个方法：public void setRequestMethod(String method) throws IOException

可以设置客户端向服务器请求的方法，可选的值有：

HttpConnection.GET：用于向服务器请求一个静态资源。GET 请求仅提供资源的 URL，不包含消息体。GET 响应用请求到的资源作为消息体。

HttpConnection.POST：用于向服务器请求一个动态服务。如果重复同一个 POST 请求可能得到不同的响应结果。POST 请求包含一个带有服务请求数据的消息体。POST 响应也包含一个响应数据的消息体。

HttpConnection.HEAD：请求与 GET 相似，只是不会在响应中返回资源。

第二个方法：public void setRequestProperty(String key, String value) throws IOException

可以用来设置普通的请求参数。例如下面的代码可以将一个命名为 Content-Language 的请求参数值设置为 en-US。

hc.setRequestProperty("Content-Language", "en-US")

完成请求参数的设置后，当输出流通过 openOutputStream()或 openDataOutputStream()方法被打开后，这两个设置请求的方法就不能再被调用，否则会引发异常。

8.2.2 获取 HTTP 连接的基本信息

HttpConnection 连接对象可以使用如下方法得到 HTTP 的基本信息：

public int getResponseCode( )throws IOException;//得到响应代码。

public String getResponseMessage( )throws IOException;//得到响应消息。

public String getProtocol( );//得到连接协议名称。

public String getHost( );//得到主机名称。

public int getPort( );//得到主机端口号。

public String getURL( );//得到请求的统一资源地址 URL。

public String getQuery( );//得到 URL 中的查询部分。

现在我们通过一个完整示例程序学习获取有关信息方法的使用。

【示例程序 8-1】

示例程序由两个文件组成，在服务器端首先运行一个 JSP 文件：文件名 login.jsp，在手机上运行的 MIDP 程序：文件名 GetInformation.java，源代码如下：

```java
import javax.microedition.io.Connector;
import javax.microedition.io.HttpConnection;
import javax.microedition.midlet.MIDlet;
import javax.microedition.midlet.MIDletStateChangeException;

public class GetInformation extends MIDlet {
    protected void startApp() throws MIDletStateChangeException {
        try{
            //连接到 HTTP 服务器端的程序 login.jsp
            HttpConnection hc =(HttpConnection)Connector.open("http://127.0.0.1:8066/mobile/login.jsp?Name=mark&PSW=123456");
```

```
            System.out.println("响应代码：" + hc.getResponseCode());
            System.out.println("响应消息：" + hc.getResponseMessage());
            System.out.println("主机：" + hc.getHost());
            System.out.println("端口：" + hc.getPort());
            System.out.println("协议：" + hc.getProtocol());
            System.out.println("URL：" + hc.getURL());
            System.out.println("查询字符串：" + hc.getQuery());
            System.out.println("请求方法：" + hc.getRequestMethod());
        }catch(Exception ex){
                ex.printStackTrace();
            }
        }
    }
    protected void destroyApp(boolean arg0) throws
                        MIDletStateChangeException {}
    protected void pauseApp() {}
}
```

本书使用 Tomcat 7.02 版本，Http 协议，端口号为 8066，在 Tomcat 服务器端指定在虚拟目录 mobile 下放置一个 login.jsp 文件，源代码如下：

```
<%@ page language="java"    contentType="text/plain; charset=GBK" %>
<%! String method, name, password; %>
<%
method = request.getMethod();
name = request.getParameter("Name");
password = request.getParameter("PSW");
out.println("welcom to the JSP");
out.println("The method you used is " + method);
out.println("Your name is " + name);
out.println("your password is " + password);
%>
```

启动 Tomcat 后，运行这个 GetInformation.java 会出现一个手机画面，在控制台上会打印出以下信息，如图 8-3 所示。

```
Running with storage root C:\Documents and Settings\Administrator\j2mewtk\2.5.2\
Running with locale: Chinese_People's Republic of China.936
Running in the identified_third_party security domain
响应代码: 200
响应消息: OK
主机: 127.0.0.1
端口: 8066
协议: http
URL: http://127.0.0.1:8066/mobile/login.jsp?Name=mark&PSW=123456
查询字符串: Name=mark&PSW=123456
请求方法: GET
```

图 8-3 获取 HTTP 连接信息

从图 8-3 可以看到，手机端通过 HttpConnection 对象可以从服务器获得响应代码、响应消息、主机地址、端口号、通讯协议、统一资源地址、查询字符串和请求方法等信息。

8.2.3 手机客户端与 HTTP 服务器通信

HTTP 编程的下一步就是作为客户端的手机与服务器端通信，所谓通信就是读和写，如果将数据发送给对方，称之为写，用到输出流；反之，从对方得到数据称为读，用到输入流。

HttpConnection 从 InputConnection 中继承了两个方法：

（1）打开输入流，返回 InputStream：

public InputStream openInputStream( ) throws IOException

（2）打开数据输入流，返回 DataInputStream：

public DataInputStream openDataInputStream( ) throws IOException

尤其第二个方法返回 DataInputStream 对象，有更强的读入能力。DataInputStream 对象可以使用：

public final int read(byte [] b) throws IOException

将对方数据以字节数组的方式读入。值得注意的是：在读取字节数组时需要知道字节数组的长度。通过查找 API 我们发现 HttpConnection 从 ContentConnection 内容连接中继承了方法：

public long getLengh()

该方法可以读到字节数组的长度。

类似的，HttpConnection 从 OutputConnection 中也继承了两个方法。

（1）打开输出流，返回 OutputStream。

public OutputStream openOutputStream( ) throws IOException

（2）打开数据输出流，返回 DataOutputStream。

public DataOutputStream openDataOutputStream( ) throws IOException

尤其第二个方法返回 DataOutputStream 对象，有更强的数据写出能力。可以使用多个 write 方法，如：

public final void writeUTF(String str) throws IOException

可以向对方写出一个字符串。

【示例程序 8-2】

作为 HTTP 编程的示例程序，我们完成一个手机远程登录系统，学习 MIDP 程序与服务器上的 JSP 程序通信。可以使用数据库查询登录密码是否正确，显示是否登录成功。为了简化程序我们只要求输入的密码与账户名相同即判断登录成功。示例程序由服务器端的 JSP 程序 Httplogin.jsp 和手机客户端的 MIDP 程序 HttpLoginMIDlet.java 组成。首先完成服务器端 Httplogin.jsp 程序，源代码如下：

<%@ page language="java" contentType="text/html;charset=gb2312"%>

```
<%
    String account = request.getParameter("ACCOUNT");
    String password = request.getParameter("PASSWORD");
    if(account.equals(password)){
        out.println("登录成功");
    }
    else{
        out.println("登录失败");
    }
%>
```

手机客户端登录 HttpLoginMIDlet 程序的代码如下:

```
import java.io.DataInputStream;
import javax.microedition.io.Connector;
import javax.microedition.io.HttpConnection;
import javax.microedition.midlet.*;
import javax.microedition.midlet.MIDlet;
import javax.microedition.midlet.MIDletStateChangeException;

public class HttpLoginMIDlet extends MIDlet implements CommandListener{
    private Form frm = new Form("HTTP 测试");
    private TextField tfAcc = new TextField("输入账号","",10, TextField.ANY);
    private TextField tfPass = new TextField("输入密码","",10,TextField.PASSWORD);
    private Command cmdLogin = new Command("登录",Command.SCREEN,1);
    private StringItem str = new StringItem("","");
    private Display dis;

    protected void startApp() throws MIDletStateChangeException {
        dis = Display.getDisplay(this);
        dis.setCurrent(frm);
        frm.append(tfAcc);
        frm.append(tfPass);
        frm.addCommand(cmdLogin);
        frm.append(str);
        frm.setCommandListener(this);
    }
    public void commandAction(Command c,Displayable d){
        if(c==cmdLogin){
            ValidateTehread vt = new ValidateTehread();
            vt.start();
        }
```

```
        }
        class ValidateTehread extends Thread{
            public void run(){
                try{
                    String url = "http://localhost:8066/mobile/Httplogin.jsp?ACCOUNT="+tfAcc.getString()+
                                 "&PASSWORD="+tfPass.getString();
                    //连接到 HTTP 服务器
                    HttpConnection hc =  (HttpConnection)Connector.open(url);
                    DataInputStream dis = hc.openDataInputStream();
                    byte[] b = new byte[(int)hc.getLength()];
                    dis.read(b);
                    if(new String(b).trim().equals("登录成功"))
                    {
                        str.setText("登录成功");
                        frm.removeCommand(cmdLogin);
                    }
                    else
                    {
                        str.setText(str.getText()+"\n 登录失败！");
                    }
                }catch(Exception ex){
                    ex.printStackTrace();
                }
            }
        }
        protected void destroyApp(boolean arg0) throws MIDletStateChangeException {}
        protected void pauseApp() {}
}
```

说明：为避免主程序被阻塞，我们将和服务器通信的程序单独写入一个线程中：class ValidateTehread extends Thread。主程序只需调用这个线程：ValidateTehread vt = new ValidateTehread(); vt.start();即可。

当我们输入相同的 ACCOUNT 和 PASSWORD 后运行结果如图 8-4 所示。

【示例程序 8-3】

手机的 MIDP 程序除要和 JSP 程序通信外，还经常要和服务器上的 Servlet 程序通信。下面通过示例程序介绍手机如何与 Servlet 程序通信。示例程序由 3 个文件组成：首先要运行服务器上的 Servlet，程序名 TestServer.java，经编译后部署于虚目录 mobile\WEB-INF\classes 子目录中，部署文件 web.xml 置于虚目录 mobile\WEB-INF 子目录下，手机上运行的 MIDP 程序 Servlet_MIDP.java。

图 8-4 登录程序运行结果

Servlet_MIDP.java 程序如下：

```java
import java.io.IOException;
import java.io.InputStream;
import java.io.OutputStream;
import javax.microedition.io.Connector;
import javax.microedition.io.HttpConnection;
import javax.microedition.lcdui.Choice;
import javax.microedition.lcdui.ChoiceGroup;
import javax.microedition.lcdui.Command;
import javax.microedition.lcdui.CommandListener;
import javax.microedition.lcdui.Display;
import javax.microedition.lcdui.Displayable;
import javax.microedition.lcdui.Form;
import javax.microedition.midlet.MIDlet;
import javax.microedition.midlet.MIDletStateChangeException;
public class Servlet_MIDP extends MIDlet implements CommandListener{
    //默认的连接地址
```

```java
private String defaultURL = "http://127.0.0.1:8066/mobile/TestServer";
//请求方法数组
private String[] methodName = {"POST","GET"};
private byte[] data=new byte[100];
//请求方法选择框
private ChoiceGroup cgMethod = new ChoiceGroup("Method:", Choice.POPUP, methodName , null);
//发送命令
private Command connectCommand = new Command ("Connect", Command.OK, 1);
//退出命令
private Command exitCommand = new Command ("Exit", Command.EXIT, 1);
private Form mainForm = new Form("HTTP Connection");
private Display display = null;
private ConnectionThread connThread = null;
private HttpConnection httpConn = null;

public Servlet_MIDP() {
    //TODO Auto-generated constructor stub
    display = Display.getDisplay(this);
    mainForm.append("URL: \n" + defaultURL);

    mainForm.append(cgMethod);
    mainForm.addCommand(connectCommand);
    mainForm.addCommand(exitCommand);
    mainForm.setCommandListener(this);
    display.setCurrent(mainForm);
}
protected void destroyApp(boolean arg0) throws MIDletStateChangeException {
    //TODO Auto-generated method stub
}
protected void pauseApp() {
    //TODO Auto-generated method stub
}
protected void startApp() throws MIDletStateChangeException {
    //TODO Auto-generated method stub
}
public void commandAction(Command c, Displayable d) {
    //TODO Auto-generated method stub
    if (c == connectCommand){
        if (connThread == null) {
```

```java
                    connThread = new ConnectionThread();
                }
                new Thread(connThread).start();
        } else if (c == exitCommand) {
                notifyDestroyed();
        }
    }

    class ConnectionThread implements Runnable{

        public void run() {
            //TODO Auto-generated method stub
            try {
                httpConn = (HttpConnection) Connector.open(defaultURL);

                //获取当前选中的请求方法
                int method = cgMethod.getSelectedIndex();
                if(method == 0) {
                    httpPostMethod();
                } else if(method == 1) {
                    httpGetMethod();
                }
            } catch (IOException ex) {
                ex.printStackTrace();
            } catch (Exception e) {
                e.printStackTrace();
            } finally {
                try {
                    if(httpConn != null) {
                        httpConn.close();
                        httpConn = null;
                    }
                } catch (Exception e) {
                    e.printStackTrace();
                }
            }
        }

        private void httpPostMethod() throws IOException{
```

```java
//TODO Auto-generated method stub
InputStream is = null;
OutputStream os = null;

try {
    //设置请求方法为 POST
    httpConn.setRequestMethod(HttpConnection.POST);

    //设置两个请求参数
    httpConn.setRequestProperty("Book_Name", "Thinking in Java");
    httpConn.setRequestProperty("Book_Language", "Chinese");

    //设置请求内容
    String reqContext = "Kehai publishing company";
    os = httpConn.openOutputStream();
    os.write(reqContext.getBytes());
    os.flush();

    int responseCode = httpConn.getResponseCode();

    //获取响应码,如果有异常则抛出
    if(responseCode != HttpConnection.HTTP_OK) {
        throw new IOException("HTTP response code: " + responseCode);
    }
    //打开一个输入流
    is = httpConn.openInputStream();
    //获取响应包数据的长度
    int len = (int)httpConn.getLength();
    if (len > 0) {
        int actual = 0;
        int bytesread = 0 ;
        byte[] data = new byte[len];
        while ((bytesread != len) && (actual != -1)) {
            actual = is.read(data, bytesread, len - bytesread);
            bytesread += actual;
        }
        //在控制台显示读取的数据
        System.out.println(new String(data));
        //在手机屏幕上显示读取的数据
```

```java
                    mainForm.append(new String(data));
                } else {
                    int ch;
                    StringBuffer sb = new StringBuffer();
                    while ((ch = is.read()) != -1) {
                        sb.append((char)ch);
                    }

                    //在控制台显示读取的数据
                    System.out.println(sb.toString());
                    //在手机屏幕上显示读取的数据
                    mainForm.append(sb.toString());
                }
        } catch (Exception e) {
            e.printStackTrace();
        } finally {
            display.setCurrent(mainForm);
            try {
                    if(httpConn != null) {
                            httpConn.close();
                            httpConn = null;
                    }
                    if(is != null) {
                            is.close();
                            is = null;
                    }
            } catch (Exception e) {
                    e.printStackTrace();
            }
        }
    }

    private void httpGetMethod() throws IOException {
        InputStream is = null;
        int responseCode = httpConn.getResponseCode();

        try {
                //获取响应码,如果有异常则抛出
                if(responseCode != HttpConnection.HTTP_OK) {
```

```java
            throw new IOException("HTTP response code: " + responseCode);
        }

        //打开一个输入流
        is = httpConn.openInputStream();

        //获取响应包数据的长度
        int len = (int)httpConn.getLength();

        if (len > 0) {
            int actual = 0;
            int bytesread = 0 ;
            byte[] data = new byte[len];
            while ((bytesread != len) && (actual != -1)) {
                actual = is.read(data, bytesread, len - bytesread);
                bytesread += actual;
            }

            //在控制台显示读取的数据
            System.out.println(new String(data));
            //在手机屏幕上显示读取的数据
            mainForm.append(new String(data));
        } else {
            int ch;
            StringBuffer sb = new StringBuffer();
            while ((ch = is.read()) != -1) {
                sb.append((char)ch);
            }

            //在控制台显示读取的数据
            System.out.println(sb.toString());
            //在手机屏幕上显示读取的数据
            mainForm.append(sb.toString());
        }
    } catch (Exception e) {
        e.printStackTrace();
    } finally {
        display.setCurrent(mainForm);
        try {
```

```java
                    if(httpConn != null) {
                            httpConn.close();
                            httpConn = null;
                    }
                    if(is != null) {
                            is.close();
                            is = null;
                    }
            } catch (Exception e) {
                    e.printStackTrace();
            }
        }
    }
}
```

TestServer.java,源程序如下:

```java
import javax.servlet.http.HttpServlet;
import java.io.IOException;
import java.io.InputStream;
import java.io.OutputStream;
import java.io.PrintWriter;
import java.util.Calendar;
import javax.servlet.ServletException;
import javax.servlet.http.HttpServletRequest;
import javax.servlet.http.HttpServletResponse;

public class TestServer extends HttpServlet {
    private OutputStream os=null;
    private InputStream is = null;
    //处理 Get 请求
    protected void doGet(HttpServletRequest request, HttpServletResponse response)
    throws ServletException, IOException {

        String respContext = "The server has recived a Get request.";
        //获取响应的输入流对象
        try{
            //获取当前时间
            Calendar calendar = Calendar.getInstance();
            String currentTime = calendar.getTime().toString();
```

```java
            //构建响应内容
            respContext = currentTime + ":" +respContext;
            os = response.getOutputStream();
            os.write(respContext.getBytes());
            os.flush();
    } catch (Exception ex) {
            ex.printStackTrace();
    } finally {
        try {
            if(os != null) {
                os.close();
                os = null;
            }
        } catch (Exception e) {
            e.printStackTrace();
        }
    }
}

//处理 Post 请求
protected void doPost(HttpServletRequest request, HttpServletResponse response)
throws ServletException, IOException {

    String respContext = "The server has recived a Post request.";
    //获取响应的输入流对象
    try{
            //获取请求参数
            String BName = request.getHeader("Book_Name");
            String BLanguage = request.getHeader("Book_Language");

            is = request.getInputStream();
            int ch;
            StringBuffer sb = new StringBuffer();
            while ((ch = is.read()) != -1) {
            sb.append((char)ch);
        }

            //获取当前时间
            Calendar calendar = Calendar.getInstance();
            String currentTime = calendar.getTime().toString();
```

```java
            //构建响应内容
            respContext = currentTime + ":" +respContext + "\n";
            respContext = respContext + "Book_Name:" + BName + "\n";
            respContext = respContext + "Book_Language:" + BLanguage + "\n";
            respContext = respContext + "Book_Pubulisher:" + sb.toString() + "\n";
            os = response.getOutputStream();
            os.write(respContext.getBytes());
            os.flush();
        } catch (Exception ex) {
                ex.printStackTrace();
        } finally {
            try {
                if(os != null) {
                    os.close();
                    os = null;
                }
            } catch (Exception e) {
                e.printStackTrace();
            }
        }
    }
}
```

部署文件 web.xml 源代码如下：

```xml
<?xml version="1.0" encoding="ISO-8859-1"?>
<web-app>
    <servlet>
     <servlet-name>TestServer</servlet-name>
     <servlet-class>TestServer</servlet-class>
    </servlet>
    <servlet-mapping>
      <servlet-name>TestServer</servlet-name>
      <url-pattern>/TestServer</url-pattern>
    </servlet-mapping>
</web-app>
```

示例程序选择传输方法界面如图 8-5 所示，选择 doPOST 方法运行结果如图 8-6 所示，选择 doGET 方法运行结果如图 8-7 所示。

网络编程　第 8 章

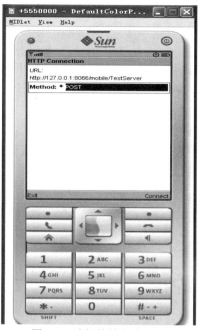

图 8-5　选择传输方法界面

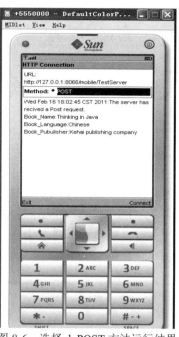

图 8-6　选择 doPOST 方法运行结果

图 8-7　选择 doGET 方法运行结果

8.3 Socket 套接字编程

从 MIDP 2.0 开始，系统提供了一些低级别网络接口的实现，低级别 IP 联网包括套接字、数据包和文件 IO 通信等。本节将讲述 Socket 协议套接字连接方式，它是基于 TCP 协议的安全连接。

本节介绍 MIDP 中的网络套接字编程方法，用到的主要 API 包是：

Javax.microedition.io

解决的基本问题与 HTTP 编程相同，同样是作为客户端的手机与服务器通信。

服务器接受客户端用到 API：

Javax.microedition.io.ServerSocket.Connection

客户端与服务器通信用到 API：

Javax.microedition.io.Socket.Connection

创建套接字连接对象与 Http 连接方法相同，同样使用 Connector 类的 open()方法创建：

public static Connection open(String name) throws IOException

8.3.1 客户端与服务器的套接字连接

（1）客户端手机连接服务器时其参数 name 打开套接字连接字符串格式为：

socket://IP 地址:端口号

如客户端要连接到 IP 地址（即服务器 IP 地址）为 218.197.118.80，端口号为 9999 的端口，返回 SocketConnection 连接对象 sc 的代码：

SocketConnection sc=(SocketConnection)Connector.open("socket://218.197.118.80:9999")

（2）服务器端监听某个端口时的连接对象类型为 serverSocketConnection，字符串格式为：

socket://:端口号

如服务器要获得监听 9999 号端口的连接对象，创建的连接对象为 ssc，则相应代码：

ServerSocketConnection ssc=(ServerSocketConnection) Connctor.open("socket://:9999")

客户端使用 SocketConnection 对象向服务器提出连接的请求，对于服务器来说应该得到客户端的这个 SocketConnection 对象，并以此为基础完成通信。获得客户端 SocketConnection 对象的任务是由 ServerSocketConnection 完成的，它从父接口 javax.microedition.io.StreamConnectionNotifier 中继承了方法：

Public SteamConnection acceptAndOpen()throws IOException

由于 SocketConnection 刚好是 StreamConnection 的子接口，所以服务器端可以使用如下代码获得连接的客户端连接对象。

SocketConnection sc=(SocketConnection)ssc.acceptAndOpen( )

8.3.2 套接字连接可以得到的基本信息

（1）通过客户端连接 SocketConnection 对象可以得到远程 IP 地址，方法如下：
public String getAddress() throws IOException
（2）得到本地地址的方法：
public String getLocalAddress() throws IOException
此外，SocketConnection 接口还提供以下两个方法设置和获取 Socket 连接的参数：
public void setSocketOption(byte option, int value) throws IllegalArgumentException, IOException
设置套接字选项的值。将标识符 option 指定的选项的值设为 value。

public int getSocketOption(byte option) throws IllegalArgumentException,IOException
返回套接字选项的值，如果值无效则返回-1。参数 option 为选项标识符。
在这两个方法中，选项参数 option 可取的值和含义如下：
DELAY：写入延迟时间。设为 0 表示禁用这个特性，非 0 表示启用该特性。
LINGER：空闲等待时间。即关闭连接前，将等待发送的数据发送完毕所需的时间，单位是 s，设为 0 时表示禁用该特性。
KEEPALIVE：Socket 连接保持连接存活的时间。设为 0 表示禁用，非 0 则启用。
RCVBUF：接收缓冲区大小，单位为字节。
SNDBUF：发送缓冲区大小，单位为字节。

【示例程序 8-4】
在上述知识的基础上，我们编写一个服务器和手机的客户端建立套接字连接的示例程序，示例程序由两个文件组成：服务器端程序文件 SocketServerMIDlet.java 和客户端程序文件 SocketClientMIDlet.java。
服务器端程序文件 SocketServerMIDlet.java，源代码如下：

```
import javax.microedition.io.Connector;
import javax.microedition.io.ServerSocketConnection;
import javax.microedition.io.SocketConnection;
import javax.microedition.lcdui.Display;
import javax.microedition.lcdui.Form;
import javax.microedition.midlet.MIDlet;
import javax.microedition.midlet.MIDletStateChangeException;

public class SocketServerMIDlet extends MIDlet {
    private Display dis;
    private Form frm = new Form("服务器端，目前未见连接");
    protected void startApp() throws MIDletStateChangeException {
        dis = Display.getDisplay(this);
        dis.setCurrent(frm);
```

```
            try{
                    ServerSocketConnection ssc = 
                            (ServerSocketConnection)Connector.open("socket://:9999");
                    SocketConnection sc = (SocketConnection)ssc.acceptAndOpen();
                    String remote = sc.getAddress();
                    frm.setTitle("服务器端，目前有" + remote + "连接上");

            }catch(Exception ex){
                    ex.printStackTrace();
            }
    }
    protected void destroyApp(boolean arg0) throws MIDletStateChangeException {}
    protected void pauseApp() {}
}
```

先运行服务器端程序，然后运行客户端程序 SocketClientMIDlet.java。源代码如下：

```
import javax.microedition.io.Connector;
import javax.microedition.io.SocketConnection;
import javax.microedition.lcdui.Command;
import javax.microedition.lcdui.CommandListener;
import javax.microedition.lcdui.Display;
import javax.microedition.lcdui.Displayable;
import javax.microedition.lcdui.Form;
import javax.microedition.midlet.MIDlet;
import javax.microedition.midlet.MIDletStateChangeException;

public class SocketClientMIDlet extends MIDlet implements CommandListener,Runnable{
    private Display dis;
    private Form frm = new Form("客户端");
    private Command cmd = new Command("连接",Command.SCREEN,1);
    protected void startApp() throws MIDletStateChangeException {
        dis = Display.getDisplay(this);
        dis.setCurrent(frm);
        frm.addCommand(cmd);
        frm.setCommandListener(this);
    }
    public void commandAction(Command arg0, Displayable arg1) {
        new Thread(this).start();
    }
    public void run(){
        try{
```

```
            SocketConnection sc = (SocketConnection)Connector.open("socket://127.0.0.1:9999");
            //连接到服务器端
            frm.setTitle("恭喜您，已经连上");
            frm.removeCommand(cmd);
        }catch(Exception ex){
            ex.printStackTrace();
        }
    }
    protected void destroyApp(boolean arg0) throws MIDletStateChangeException {}
    protected void pauseApp() {}
}
```

示例程序套接字连接运行结果如图 8-8 和图 8-9 所示。

图 8-8 （服务器）套接字连接服务器端

图 8-9 （客户端）套接字连接服务器端

8.3.3 套接字连接通信

使用套接字通信同样是利用输入输出流。首先介绍打开输入输出流的方法：

1. 打开输入流

SocketConnection 同 HTTPConnection 一样，从 InputConnection 中继承了两个创建输入流对象的方法：

（1）创建输入流，返回 InputStream：
public InputStream openInputStream() throws IOException
（2）创建数据输入流，返回 DataInputStream：
public DataInputStream openDataInputStream() throws IOException
第二个方法返回 DataInputStream 对象，有更强的读入能力，数据输入流对象可以使用：
public final int read(byte [] b) throws IOException
将对方数据以字节数组的方式读入。

并且，DataInputStream 对象可以直接读取一个字符串：
Public final String readUTF() throws IOEception
2．打开输出流
SocketConnection 同样从 OutputConnection 中继承了两个创建输出流对象的方法：
public OutputStream() throws IOEception
public DataOutputStream openDataOutputStream() throws IOException
第二个方法返回 DataOutputStream 对象，有更强的写出能力，可以使用多个 write 方法。如：
public final void write(byte [] b, int off, int len) throws IOEception
可以向对方写出一个字节数组：
public final void writeUTF(String str) throws IOException
还可以向对方直接写出一个字符串。

【示例程序 8-5】
利用这些基础知识，我们完成一个用套接字实现服务器与客户端双向聊天的示例程序。
服务器端程序：TCPServerMIDlet.java

```java
import java.io.DataInputStream;
import java.io.DataOutputStream;
import javax.microedition.io.Connector;
import javax.microedition.io.ServerSocketConnection;
import javax.microedition.io.SocketConnection;
import javax.microedition.lcdui.Command;
import javax.microedition.lcdui.CommandListener;
import javax.microedition.lcdui.Display;
import javax.microedition.lcdui.Displayable;
import javax.microedition.lcdui.Form;
import javax.microedition.lcdui.TextField;
import javax.microedition.midlet.MIDlet;
import javax.microedition.midlet.MIDletStateChangeException;

public class TCPServerMIDlet extends MIDlet implements CommandListener{
    private ServerSocketConnection ssc = null;
    private SocketConnection sc = null;
    private DataInputStream dis = null;
    private DataOutputStream dos = null;
```

```java
private TextField tfMsg = new TextField("输入聊天信息","",255,TextField.ANY);
private Command cmdSend = new Command("发送",Command.SCREEN,1);
private Form frmChat = new Form("聊天界面：服务器端");
private Display display;

protected void startApp() throws MIDletStateChangeException {
    display = Display.getDisplay(this);
    display.setCurrent(frmChat);
    frmChat.addCommand(cmdSend);
    frmChat.append(tfMsg);
    frmChat.setCommandListener(this);
    frmChat.append("以下是聊天记录：\n");
    try{
        ssc = (ServerSocketConnection)Connector.open("socket://:9999");
        sc = (SocketConnection)ssc.acceptAndOpen();
        dis = sc.openDataInputStream();
        dos = sc.openDataOutputStream();
        new ReceiveThread().start();
    }catch(Exception ex){
        ex.printStackTrace();
    }
}
public void commandAction(Command c,Displayable d){
    if(c==cmdSend){
        try{
            dos.writeUTF("服务器说：" + tfMsg.getString());
        }catch(Exception ex){}
    }
}
class ReceiveThread extends Thread{
    public void run(){
        while(true){
            try{
                String msg = dis.readUTF();
                frmChat.append(msg + "\n");
            }catch(Exception ex){ex.printStackTrace();}
        }
    }
}
protected void destroyApp(boolean arg0) throws MIDletStateChangeException {}
protected void pauseApp() {}
}
```

客户端程序：TCPGroupClientMIDlet.java

```java
import java.io.DataInputStream;
import java.io.DataOutputStream;
import javax.microedition.io.Connector;
import javax.microedition.io.SocketConnection;
import javax.microedition.lcdui.*;
import javax.microedition.midlet.*;

public class TCPClientMIDlet extends MIDlet implements CommandListener{
    private SocketConnection sc = null;
    private DataInputStream dis = null;
    private DataOutputStream dos = null;
    private TextField tfMsg = new TextField("输入聊天信息","",255,TextField.ANY);
    private Command cmdSend = new Command("发送",Command.SCREEN,1);
    private Form frmChat = new Form("聊天界面：客户端");
    private Display display;
    protected void startApp() throws MIDletStateChangeException {
        display = Display.getDisplay(this);
        display.setCurrent(frmChat);
        frmChat.addCommand(cmdSend);
        frmChat.append(tfMsg);
        frmChat.setCommandListener(this);
        frmChat.append("以下是聊天记录：\n");
        try{
            sc = (SocketConnection)Connector.open("socket://127.0.0.1:9999");
            dis = sc.openDataInputStream();
            dos = sc.openDataOutputStream();
            new ReceiveThread().start();
        }catch(Exception ex){
            ex.printStackTrace();
        }
    }
    public void commandAction(Command c,Displayable d){
        if(c==cmdSend){
            try{
                dos.writeUTF("客户端说： " + tfMsg.getString());
            }catch(Exception ex){}
        }
    }
    class ReceiveThread extends Thread{
```

```
        public void run(){
            while(true){
                try{
                    String msg = dis.readUTF();
                    frmChat.append(msg + "\n");
                }catch(Exception ex){}
            }
        }
    }
    protected void destroyApp(boolean arg0) throws MIDletStateChangeException {}
    protected void pauseApp() {}
}
```

示例程序服务器端运行结果如图 8-10 所示,客户端运行结果如图 8-11 所示。

图 8-10 服务器端运行结果

图 8-11 客户端运行结果

【示例程序 8-6】

这个程序只需稍加修改就可以成为实现多客户端群聊功能的示例程序。

服务器端程序:TCPGroupServerMIDlet.java

```
import java.io.DataInputStream;
import java.io.DataOutputStream;
import java.util.Vector;
```

```java
import javax.microedition.io.Connector;
import javax.microedition.io.ServerSocketConnection;
import javax.microedition.io.SocketConnection;
import javax.microedition.midlet.MIDlet;
import javax.microedition.midlet.MIDletStateChangeException;

public class TCPGroupServerMIDlet extends MIDlet implements Runnable{

    private ServerSocketConnection ssc = null;
    private SocketConnection sc = null;
    private Vector clients = new Vector();
    private boolean CANACCPT = true;
    protected void startApp() throws MIDletStateChangeException {
        try{
            ssc = (ServerSocketConnection)Connector.open("socket://:9999");
            new Thread(this).start();
        }catch(Exception ex){
            ex.printStackTrace();
        }
    }
    public void run(){
        while(CANACCPT){//不断接受客户端连接
            try{
                sc = (SocketConnection)ssc.acceptAndOpen();
                //开一个线程给这个客户端
                ChatThread ct = new ChatThread(sc);
                clients.addElement(ct);//将线程添加进集合
                ct.start();
            }catch(Exception ex){
                ex.printStackTrace();
            }
        }
    }
    protected void destroyApp(boolean arg0) throws MIDletStateChangeException {}
    protected void pauseApp() {}
    //每连接上一个客户端，就开一个聊天线程
    class ChatThread extends Thread{
        private DataInputStream dis;
        private DataOutputStream dos;
        private boolean CANREAD = true;
        public ChatThread(SocketConnection sc){
```

```java
            try{
                dis = sc.openDataInputStream();
                dos = sc.openDataOutputStream();
            }catch(Exception ex){
                ex.printStackTrace();
            }
        }
        //负责读取相应 SocketConnection 的信息
        public void run(){
            while(CANREAD){
                try{
                    String str = dis.readUTF();
                    //将该信息发送给所有客户端，访问集合中的所有线程
                    for(int i=0;i<clients.size();i++){
                        ChatThread ct = (ChatThread)clients.elementAt(i);
                        ct.dos.writeUTF(str);
                    }

                }catch(Exception ex){
                    ex.printStackTrace();
                }
            }
        }
    }
}
```

客户端程序：TCPGroupClientMIDlet.java

```java
import java.io.DataInputStream;
import java.io.DataOutputStream;
import javax.microedition.io.*;
import javax.microedition.lcdui.*;
import javax.microedition.midlet.*;

public class TCPGroupClientMIDlet extends MIDlet implements CommandListener,Runnable{
    private SocketConnection sc = null;
    private DataInputStream dis = null;
    private DataOutputStream dos = null;
    private boolean ISRUN = true;
    private TextField tfNickName = new TextField("输入昵称","",10,TextField.ANY);
    private TextField tfMsg = new TextField("输入聊天信息","",255,TextField.ANY);
    private Command cmdSend = new Command("发送",Command.SCREEN,1);
```

```java
        private Form frmChat = new Form("客户端聊天界面");
        private Display display;

        protected void startApp() throws MIDletStateChangeException {
            display = Display.getDisplay(this);
            display.setCurrent(frmChat);
            frmChat.append(tfNickName);
            frmChat.append(tfMsg);
            frmChat.addCommand(cmdSend);
            frmChat.setCommandListener(this);
            frmChat.append("以下是聊天记录：\n");
            try{
                sc=(SocketConnection)Connector.open("socket://127.0.0.1:9999");
                dis = sc.openDataInputStream();
                dos = sc.openDataOutputStream();
                new Thread(this).start();
            }catch(Exception ex){
                ex.printStackTrace();
            }
        }
        public void commandAction(Command c,Displayable d){
            try{
                dos.writeUTF(tfNickName.getString() + "说：" + tfMsg.getString());
            }catch(Exception ex){
                ex.printStackTrace();
            }
        }
        public void run(){
            while(ISRUN){
                try{
                    String msg = dis.readUTF();
                    frmChat.append(msg + "\n");
                }catch(Exception ex){}
            }
        }
    protected void destroyApp(boolean arg0) throws MIDletStateChangeException {}
        protected void pauseApp() {}
}
```

服务器端运行结果为空白，昵称为"客户端AA"的运行结果如图 8-12 所示，昵称为"客户端B"的运行结果如图 8-13 所示。

图 8-12　客户端 AA 运行图　　　　图 8-13　客户端 B 运行图

8.4　UDP 数据报编程

在此之前介绍的套接字接口是基于 TCP 协议的安全连接，是面向连接的类型。与 TCP 同属 TCP/IP 协议体系的传输层协议还有 UDP 协议，而 Datagram 和 DatagramConnection 接口是基于 UDP 数据报、面向无连接的数据类型，只负责传输信息，并不保证信息一定会被收到，虽然安全性稍差，但传输速度较快。所使用的协议是 UDP 协议，采用同一协议的还有 DNS、SNMP、QQ 等，使用网络编程 API 是：

javax.microedition.io.UDPDatagramConnection

8.4.1　客户端与服务器端数据报连接

UDP 数据报通信方式仍然是手机作为客户端将信息发到服务器，然后由服务器转发到目标客户端。所以通信的客户端必须首先连接到服务器的 IP 地址和端口，在服务器端则必须监听某个端口。

在 UDP 数据报编程中服务器端对端口的监听是由 UDPDatagramConnection 类的对象完成的。数据报的这个连接对象仍然需要使用 Connector 类的 open() 方法得到。

创建服务器端连接作为 open() 方法参数的字符串的格式为：

datagram://:端口号

如 datagram://:9999 表示数据报服务器监听 9999 号端口。

如下代码可以获得服务器上监听 9999 号端口的连接对象 udc

UDPDatagramConnection udc=(UDPDataConnection)Connector.open("datagram://:9999");

对于客户端要连接到服务器，连接作为 open()方法参数的字符串的格式：

datagram://服务器 IP:端口号

客户端希望连接到本机服务器获得数据报连接对象的方法如下：

udc=(UDPDataConnection)Connector.open("datagram://127.0.0.1:9999");

对于面向连接的 TCP 套接字通信，服务器需要 acceptAndOpen()方法等待客户端的连接；而对于无连接的数据报通信则是不需要的。

8.4.2 数据包的传递

在套接字通信中要用到输入输出流，然而在 UDP 情况下则是采用另外一种数据包的形式进行。发送数据包为输出，接收数据包为输入。为此 UDPDatagramConnection 从 DatagramConnection 中继承了两个发送和接收数据包的方法：

（1）接收数据包：

public void receive(Datagram dgram)throws IOException

（2）发送数据包：

public void send(Datagram dgram)throws IOException

这两个方法的参数 Datagram 是个接口，它同时具有输入和输出功能。由于是接口，则不能用构造方法创建对象。

Datagram 对象是由 UDPDatagramConnection 从 DatagramConnection 继承了下列方法创建的：

public Datagram newDatagram(int size)throws IOException;

public Datagram newDatagram(byte[] buf,int size)throws IOException;

public Datagram newDatagram(byte [] buf, int size , String addr)throws IOException;

public Datagram newDatagram(int size,String addr)throws IOException;

参数 int size 指定 Datagram 所需数据缓冲区的大小，byte[] buf 为数据报内部使用的缓冲区用于装载数据包内容，String addr 为数据报发送或接收地址。

数据报一定有发送到的地址才知道发送到哪里，从创建方法可以看到有的数据报是没有发送地址的。那么如何指定发送地址？规律如下：

（1）客户端在确定服务器的情况下，所创建的 Datagram 对象不需要设置地址，数据包可直接发送到服务器。

（2）服务器事先不知道客户端的 IP 地址，服务器可以使用如下方法将另一个数据报的地址设定为发送地址：

public void setAddress(Datagram reference);

由此可以知道一般通信时，客户端首先给服务器发送一个 Datagram，让服务器用这个数据报作为参考，知道客户端地址，之后与客户端通信。

Datagram 类还有一些数据报通信常用方法：

（1）设定数据报的长度：

public void setLenth(int len);

（2）以字符串的形式设定发送地址：

public void setAddress(String addr)throws IOException

（3）设置数据包的数据：

public void setData(byte [] buffer, int offset, int len);

（4）得到发送地址：

public String getAddress();

（5）得到数据：

public byte[] getData();

（6）得到数据报的长度：

public int getLenth();

【示例程序 8-7】

在这些知识的基础上我们就可以编写一个手机客户端与服务器聊天的示例程序：

首先建立服务器端程序 UDPServerMIDlet.java：

```java
import javax.microedition.io.Connector;
import javax.microedition.io.Datagram;
import javax.microedition.io.UDPDatagramConnection;
import javax.microedition.lcdui.Command;
import javax.microedition.lcdui.CommandListener;
import javax.microedition.lcdui.Display;
import javax.microedition.lcdui.Displayable;
import javax.microedition.lcdui.Form;
import javax.microedition.lcdui.TextField;
import javax.microedition.midlet.MIDlet;
import javax.microedition.midlet.MIDletStateChangeException;

public class UDPServerMIDlet extends MIDlet implements CommandListener {
    private UDPDatagramConnection udc = null;
    private String address = null;
    private TextField tfMsg = new TextField("输入聊天信息", "", 255, TextField.ANY);
    private Command cmdSend = new Command("发送", Command.SCREEN, 1);
    private Form frmChat = new Form("聊天界面：服务器端");
    private Display display;

    protected void startApp() throws MIDletStateChangeException {
        display = Display.getDisplay(this);
        display.setCurrent(frmChat);
        frmChat.addCommand(cmdSend);
```

```java
            frmChat.append(tfMsg);
            frmChat.setCommandListener(this);
            frmChat.append("以下是聊天记录：\n");
            try {
                udc = (UDPDatagramConnection) Connector.open("datagram://:9999");
                new ReceiveThread().start();
            } catch (Exception ex) {
                ex.printStackTrace();
            }
        }

    public void commandAction(Command c, Displayable d) {
        if (c == cmdSend) {
            try {
                String msg = "服务器说："+ tfMsg.getString();
                byte[] data = msg.getBytes();
                Datagram datagram = udc.newDatagram(data, data.length, address);
                udc.send(datagram);
            } catch (Exception ex) {
            }
        }
    }

    class ReceiveThread extends Thread {
        public void run() {
            while (true) {
                try {
                    Datagram datagram = udc.newDatagram(255);
                    udc.receive(datagram);
                    String msg = new String(datagram.getData(), 0, datagram.getLength());
                    frmChat.append(msg + "\n");
                    address = datagram.getAddress();
                } catch (Exception ex) {
                    ex.printStackTrace();
                }
            }
        }
    }

    protected void destroyApp(boolean arg0) throws MIDletStateChangeException {
    }
```

```java
    protected void pauseApp() {
    }
}
```

客户端运行程序：UDPClientMIDlet.java

```java
import javax.microedition.io.Connector;
import javax.microedition.io.Datagram;
import javax.microedition.io.UDPDatagramConnection;
import javax.microedition.lcdui.Command;
import javax.microedition.lcdui.CommandListener;
import javax.microedition.lcdui.Display;
import javax.microedition.lcdui.Displayable;
import javax.microedition.lcdui.Form;
import javax.microedition.lcdui.TextField;
import javax.microedition.midlet.MIDlet;
import javax.microedition.midlet.MIDletStateChangeException;

public class UDPClientMIDlet extends MIDlet implements CommandListener{
    private UDPDatagramConnection udc = null;
    private TextField tfMsg = new TextField("输入聊天信息","",255, TextField.ANY);
    private Command cmdSend = new Command("发送",Command.SCREEN,1);
    private Form frmChat = new Form("聊天界面：客户端");
    private Display display;

    protected void startApp() throws MIDletStateChangeException {
        display = Display.getDisplay(this);
        display.setCurrent(frmChat);
        frmChat.addCommand(cmdSend);
        frmChat.append(tfMsg);
        frmChat.setCommandListener(this);
        frmChat.append("以下是聊天记录：\n");
        try{
            udc = (UDPDatagramConnection)Connector.open("datagram://127.0.0.1:9999");
            new ReceiveThread().start();
        }catch(Exception ex){
            ex.printStackTrace();
        }
    }
    public void commandAction(Command c,Displayable d){
        if(c==cmdSend){
            try{
                String msg = "客户端说：" + tfMsg.getString();
                byte[] data = msg.getBytes();
                Datagram datagram = udc.newDatagram(data,data.length);
```

```
                    udc.send(datagram);
                }catch(Exception ex){}
            }
        }
        class ReceiveThread extends Thread{
            public void run(){
                while(true){
                    try{
                        Datagram datagram = udc.newDatagram(255);
                        udc.receive(datagram);
                        String msg = new String(datagram.getData(),0,datagram.getLength());
                        frmChat.append(msg + "\n");
                    }catch(Exception ex){ex.printStackTrace();}
                }
            }
        }
        protected void destroyApp(boolean arg0) throws MIDletStateChangeException {}
        protected void pauseApp() {}
}
```

服务器端运行结果如图 8-14 所示，客户端运行结果如图 8-15 所示，在运行时，要首先从客户端开始向服务器发送信息，服务器才可以回应客户端的请求，为什么？读者可以思考。

图 8-14　服务器端运行结果

图 8-15　客户端运行结果

【示例程序 8-8】

前面讲了服务器与客户端的聊天系统,但实际应用中需要的是客户端与客户端的信息交换。客户对客户的聊天系统实际上是客户端的数据报经服务器转发。下边我们实现多客户端的聊天系统示例程序。

首先写出服务器端程序 UDPGroupServerMIDlet.java:

```java
import java.util.Vector;
import javax.microedition.io.Connector;
import javax.microedition.io.Datagram;
import javax.microedition.io.UDPDatagramConnection;
import javax.microedition.midlet.MIDlet;
import javax.microedition.midlet.MIDletStateChangeException;

public class UDPServerMIDlet7_8 extends MIDlet implements Runnable{
    private UDPDatagramConnection udc = null;
    private Vector addresses = new Vector();//保存所有客户端地址
    private boolean ISRUN = true;
    protected void startApp() throws MIDletStateChangeException {
        try{
            udc = (UDPDatagramConnection)Connector.open("datagram://:9999");
            new Thread(this).start();
        }catch(Exception ex){
            ex.printStackTrace();
        }
    }
    public void run(){
        while(ISRUN){//读取信息
            try{
                Datagram datagram = udc.newDatagram(255);
                udc.receive(datagram);
                String msg = new String(datagram.getData(),0,datagram.getLength());
                //维护地址集合
                String address = datagram.getAddress();
                if(!addresses.contains(address)){
                    addresses.addElement(address);
                }
                this.sendToAll(msg.getBytes());//发送给所有客户端
            }catch(Exception ex){
                ex.printStackTrace();
```

```java
            }
        }
    }
    public void sendToAll(byte[] data) throws Exception{
        for(int i=0;i<addresses.size();i++){
            String address = (String)addresses.elementAt(i);
            Datagram datagram = udc.newDatagram(data,data.length,address);
            udc.send(datagram);
        }
    }
    protected void destroyApp(boolean arg0) throws MIDletStateChangeException {}
        protected void pauseApp() {}
}
```

再给出客户端运行程序 UDPGroupClientMIDlet.java，只需在两个项目中运行两个不同名字但内容相同的客户端程序即可：

```java
import javax.microedition.io.Connector;
import javax.microedition.io.Datagram;
import javax.microedition.io.UDPDatagramConnection;
import javax.microedition.lcdui.Command;
import javax.microedition.lcdui.CommandListener;
import javax.microedition.lcdui.Display;
import javax.microedition.lcdui.Displayable;
import javax.microedition.lcdui.Form;
import javax.microedition.lcdui.TextField;
import javax.microedition.midlet.MIDlet;
import javax.microedition.midlet.MIDletStateChangeException;

public class UDPGroupClientMIDlet extends MIDlet implements CommandListener,Runnable{

    private UDPDatagramConnection udc = null;
    private boolean ISRUN = true;
    private TextField tfNickName = new TextField("输入昵称","",10,TextField.ANY);
    private TextField tfMsg = new TextField("输入聊天信息","",255,TextField.ANY);
    private Command cmdSend = new Command("发送",Command.SCREEN,1);
    private Form frmChat = new Form("客户端聊天界面");
    private Display display;
    protected void startApp() throws MIDletStateChangeException {
        display = Display.getDisplay(this);
        display.setCurrent(frmChat);
```

```java
        frmChat.append(tfNickName);
        frmChat.append(tfMsg);
        frmChat.addCommand(cmdSend);
        frmChat.setCommandListener(this);
        frmChat.append("以下是聊天记录：\n");
        try{
            udc = (UDPDatagramConnection)Connector.open("datagram://127.0.0.1:9999");
            Datagram datagram = udc.newDatagram(0);
            udc.send(datagram);
            new Thread(this).start();
        }catch(Exception ex){
            ex.printStackTrace();
        }
    }
    public void commandAction(Command c,Displayable d){
        try{
            String msg = tfNickName.getString() + "说："  + tfMsg.getString();
            byte[] data = msg.getBytes();
            Datagram datagram = udc.newDatagram(data,data.length);
            udc.send(datagram);
        }catch(Exception ex){
            ex.printStackTrace();
        }
    }
    public void run(){
        while(ISRUN){
            try{
                Datagram datagram = udc.newDatagram(255);
                udc.receive(datagram);
                String msg = new String(datagram.getData(),0,datagram.getLength());
                frmChat.append(msg + "\n");
            }catch(Exception ex){
                ex.printStackTrace();
            }
        }
    }
    protected void destroyApp(boolean arg0) throws MIDletStateChangeException {}
    protected void pauseApp() {}
}
```

服务器端看到的结果如图 8-16 所示，客户端运行结果如图 8-17 和图 8-18 所示。

图 8-16　服务器看到画面

图 8-17　客户端运行画面-1

图 8-18　客户端 0 运行画面-2

8.5　本章小结

本章首先明确了在 Java ME 平台下进行网络开发的基本框架，然后介绍了 MIDP 对该基本框架的扩展，形成自己的网络连接类/接口体系。随后逐步细化，讲解了 HTTP 连接、套接字的连接、UDP 协议的数据报连接的概念和应用，以及手机之间使用各种协议的通信问题。

9 MMAPI 多媒体程序设计

本章主要介绍 Java ME MMAPI 多媒体方面的内容。读者需要掌握以下知识点：
- 多媒体资源的播放
- 多媒体资源的控制
- 多媒体资源的存储
- 多媒体资源的管理

9.1 移动媒体 API（MMAPI）概述

MMAPI 的特性如下：
- 支持音频和视频的播放和记录，以及音调生成。
- 较少的资源消耗。
- 支持 CLDC 平台。
- 不局限于某个特定的媒体协议。
- 可以随时对 API 进行裁减和扩展。

对于不同格式的媒体数据和传输协议，MMAPI 提供了一套规范的、可扩展的、简单的程序接口。

9.1.1 MMAPI 的体系结构

MMAPI 主要涉及四个基本概念：管理器（Manager）、播放器（Player）、控制器（Controller）和数据源（DataSource），如图 9-1 所示。

管理器：是媒体处理的总控制者，它提供了静态方法来创建播放器和测试手机所支持的媒体类型和协议。

播放器：专门实现媒体内容的播放，它提供了管理播放器生命周期及重放次数的方法。

控制器：用于实现播放器的各种控制功能的接口，如音量控制等。

数据源：是对媒体协议处理器的抽象，它负责为播放器提供相关的多媒体数据。

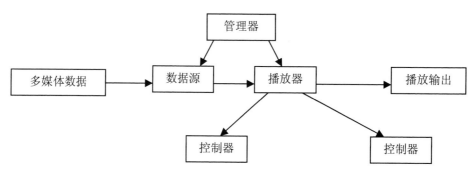

图 9-1　MMAPI 的基本体系结构

9.1.2　管理器 Manager 类

Manager 类主要提供了静态方法来创建播放器 Player 对象及测试硬件系统所支持的媒体类型和协议。

Manager 类提供了创建 Player 对象的两个重载静态方法：

```
public static Player createPlayer(String locator)
                Throws java.io.IOException,MediaException
public static Player createPlayer(InputStream stream, String type)
                Throws java.io.IOException,MediaException
```

创建 Player 对象的第一个方法：参数 locator 是指向媒体数据的定位字符串，它有下列 3 种类型：

（1）媒体文件的网络地址。可以是指向远程 HTTP 服务器媒体文件的 URL，如 http://www.meidia.com/mid/gaoshanliushui.mid。

（2）媒体数据捕获模式。该类目前只有两个值：一个是 capture//audio，用于音频截取；另一个是 capture//video，用于在手机上截取静态图片。

（3）内存中的空数据类型。用于实例化一个空播放器。之后可以用 MIDIControl（即 Musical Instrument Digital Interface Control）和 ToneControl 对象动态设置其内容。这种类型的参数也只有两个：一个是 device//midi 类型的 Manager.MIDI_DEVICE_LOCATOR，另一个是 device://tone 类型的 Manager.TONE_DEVICE_LOCATOR。

创建 Player 对象的第二种方法：是从输入流 stream 中读取数据来创建 Player 对象。其中参数类型则指定为媒体的 MIME（即 Multipurpose Internet Mail Extensions，多功能 Internet 邮件扩充服务）类型。常见的媒体类型有：

Audio/midi：代表 MIDI 文件。

audio/sp-midi：代表可扩展多和弦 MIDI。

audio/x-tone-seq：MIDP 2.0 定义的单音序列。

audio/x-wav：代表 WAV PCM 采样音频。

image/gif：代表 GIF 89a（GIF 动画）或 gif 图片格式。

video/mpeg：MPEG 视频。

video/vnd.sun.rgb565：代表视频记录格式。

Manager 类的另一个功能是提供两种方法，测试手机所支持的媒体类型和协议。

我们可以使用 getSupportedContentTypes 方法获得当前设备支持的媒体类型。该方法的定义如下：

public static String[] getSupportedContentTypes(String protocol);

返回值是表示各种媒体类型的字符串数组。对于音频，如 WAV 音频为 audio/x-wav，MP3 音频为 audio/mpeg，MIDI 音频为 audio/midi 等。参数是相关媒体的协议。对于音频有：http 协议：超文本传输，device 协议：设备协议，capture 协议：媒体捕获协议等。

相反地，Manager 类还可以由媒体类型得到设备所支持的协议，获取设备所支持的协议的方法定义如下：

public static String[] getSupportedContentTypes(String content_type);

返回值是表示各种协议的字符串数组。

此外 Manager 类还直接提供播放单音符的方法：

public static void playTone(int note, int duration, int volume);

其中 note 表示单音符，取值范围是 0～127，按照高音到低音的顺序排列；duration 表示声音持续的时间，单位是毫秒；volume 表示音量，取值范围是 0～100，按照音量从低到高的顺序排列。例如：Manager.playTone(50,1000,80);表示播放音调为 50、音量为 80 的音符，持续 1 秒钟。

9.1.3 播放器 Player 接口

Player 接口控制媒体的播放过程，主要功能是提供方法来管理播放器的生命周期。一个 Player 对象的生命周期一般有 5 种状态：未实例化、实例化、预读取、启动和关闭。

（1）未实例化（Unrealized）：Player 对象刚刚被创建，尚未分配任何资源。

（2）实例化（Realized）：进入实例化状态，表明 Player 对象已经获得了所需资源数据。此时如果调用 dealLocate()方法，就会返回未实例化状态。

（3）预读取（Prefetched）：在实例化状态下，调用 prefetch()方法就会进入预读取状态。在这个状态下，Player 对象需要完成很多启动所必需的操作（如获得扬声器或照相机的控制权等)。一旦预读取成功,就意味着播放器立刻可以启动了。如果在该状态下调用 deallocate()方法,则返回到实例化状态。

(4)启动(Started):预读取成功之后,调用 start()方法,则进入启动状态。此时播放器开始播放媒体,捕获音频或相机取景视频。启动后可以调用 stop()方法停止播放,返回预读取状态。

(5)关闭(Closed):通过 Close()方法,可以进入关闭状态。此时 Player 对象将释放全部资源,不能被再次使用。

播放器对象状态转换如图 9-2 所示。

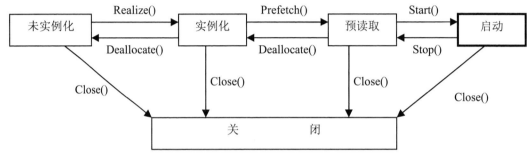

图 9-2 Player 播放器状态转换图

播放器对象可以通过 PlayerListener 接口监听播放器事件来控制对方行为。PlayerListener 接口中只有一个方法:

public void playerUpdate(Player player, String event, Object eventData)

其中参数 player 表示发出消息的播放器实例;event 表示事件的类型,如 END_OF_MEDIA 表示媒体播放事件完成一次;eventData 表示发出消息的有关数据。

Player 对象加载监听器的方法:

public void addPlayerListener(PlayerListener playerListener)

功能:为 Player 对象添加 PlayerListener 监听器。

删除监听器的方法:

public void removePlayerListener(PlayerListener playerListener)

功能:移去 PlayerListener 监听器。

此外 Player 对象还有一些方法:

public void setLoopCount(int count)

为播放器设置媒体播放循环次数的方法。

public long getMediaTime()

获取播放器的媒体时间。

还有其他方法可以查阅有关 API。

9.1.4 数据源 DataSource 类

DataSource 类是对媒体协议处理器的抽象，它隐藏了来自文件、输入流或是其他传输机制的媒体数据如何被读取的细节，并提供了 Player 对象访问媒体数据的方法。不过在实际开发中使用数据源对象较少，我们在此不再赘述。

9.1.5 控制器 Control 接口

在 MMAPI 中定义了很多控制器接口，通过这些控制器可以方便地控制媒体播放的过程。由于 Player 接口本身继承于 Controllable 接口，故 Player 对象可以直接调用 getControls()和 getControl()方法，来获取当前播放器加载的控制器。其方法定义：

```
public Contrl[] getControls();
public Contrl getControl( String controlType);
```

对于不同的手机支持的控制器是不一样的，需要程序员根据设备选择。这里举出 MMAPI 定义的 12 个控制器。

MetaDataControl：用来从媒体中获得元数据信息。
MIDIControl：提供对播放器表现和传输设备的访问。
GUIControl：代表了一个具有用户界面组件的控制操作。
PitchControl：改变重放位置，而不改变重放的速度。
RateControl：控制重放速率。
TempoControl：控制 MIDI 歌曲的节奏。
VolumeControl：控制音量。
VideoControl：控制可视内容的显示。
FramePositioningControl：可以精确定位视频的帧。
RecordControl：记录当前被 Player 播放的内容。
StopTimeControl：使应用程序为 Player 对象预订停止时间。
ToneControl：是一个可以播放用户自定义音调序列的接口。

9.2 音频播放

对于音频播放，有时需要控制音量的大小或者静音，这需要音量控制接口 VolumeControl。音量控制接口使用 0～100 作为控制的线性空间。0 代表没声，100 代表最大，但最大和最小的音量值完全依赖于具体的硬件实现。

接口中定义了设置音量的方法：

```
public void setLevel(int level);
```

参数 level 取值范围 0~100。

获得当前音量的方法：

public int getLevel();

返回值在 0~100 之间，但也可能是-1，表示当前设备还没有初始化。

具体代码可参考如下程序段：

myVolumeControl = (VolumeControl)player.getControl("VolumeControl");
        //获得播放器自身的 VolumeControl 接口
 myVolumeControl.setLevel(volume);

利用 VolumeControl 可以动态设置音量。

我们还可以使用下面方法设置静音：

public void setMute(Boolean mute);

参数 mute 取值为 true 时，表示播放器处于静音状态。

【示例程序 9-1】

下面我们举出简单音乐播放的示例程序。在音乐播放过程中，通过音量控制器 VolumeControl 改变音量的大小，通过 stop()和 Start()方法实现暂停和继续播放。如图 9-3 所示为音乐播放画面。

程序文件名：PlayMusic.java

```java
import java.io.IOException;
import java.io.InputStream;
import javax.microedition.lcdui.*;
import javax.microedition.media.*;
import javax.microedition.midlet.*;

public class PlayMusic extends MIDlet implements CommandListener {
    private Form form;
    private Display display;
    private Gauge gauge;
    private Player player;
    //标志播放器是否处于暂停状态
    private boolean isPause = false;

    //音量控制器
    private VolumeControl volumeControl;

    //当前音量值
    private int currentVolume;

    //播放命令
    private Command playCommand = new Command("Play", Command.OK, 1);
```

```java
//暂停命令
private Command pauseCommand = new Command("Pause", Command.OK, 1);

//退出命令
private Command exitCommand = new Command("Exit", Command.EXIT, 1);

public PlayMusic() {
    //TODO Auto-generated constructor stub
    form = new Form("Play Music Demo");
    display = Display.getDisplay(this);
}

protected void destroyApp(boolean arg0) throws MIDletStateChangeException {
    //关闭播放器
    if(player != null)
        player.close();
}

protected void pauseApp() {   }

protected void startApp() throws MIDletStateChangeException {
    //用 Gauge 实现一个音量调节界面，并设置初始音量为 50
    gauge = new Gauge("Volume Value", true, 100, 0);
    gauge.setValue(50);
    form.append(gauge);
    form.addCommand(playCommand);
    form.addCommand(exitCommand);
    form.setCommandListener(this);
    display.setCurrent(form);
}
public void commandAction(Command c, Displayable d) {
    if(c == playCommand) {
        if(isPause) {
            try {
                //从暂停的地方继续播放
                player.start();
            } catch (MediaException e) {
                e.printStackTrace();
            }
            isPause = false;
        } else
```

```java
            playMusic();
    form.removeCommand(playCommand);
    form.addCommand(pauseCommand);
    //启动一个线程，每隔100ms，检查音量是否改变
    new Thread() {
        public void run() {
            while(!isPause){
                int value = gauge.getValue();
                if(value != currentVolume){
                    currentVolume = value;
                    volumeControl.setLevel(currentVolume);
                }
            }
            try {
                Thread.sleep(100);
            } catch (InterruptedException e) {
                e.printStackTrace();
            }
        }
    }.start();
} else if(c == pauseCommand) {
    try {
        player.stop();
    } catch (MediaException e) {
        e.printStackTrace();
    }
    form.removeCommand(pauseCommand);
    form.addCommand(playCommand);
    isPause = true;
} else if(c == exitCommand) {
    notifyDestroyed();
}
}
/**
 * 实现播放音乐
 */
private void playMusic() {
    //读取本地媒体资源，转化为输入流对象
    InputStream is = this.getClass().getResourceAsStream("/1.mid");
    try {
        //关闭播放器
```

```java
            if(player != null)
                player.close();
            player = Manager.createPlayer(is, "audio/midi"); //构造播放器对象
            player.addPlayerListener(new PlayMusicListener());//为播放器注册事件监听
            player.realize();    //播放器序列化
            //创建一个音量控制器
            volumeControl = (VolumeControl)player.getControl("VolumeControl");
            currentVolume = gauge.getValue();
            volumeControl.setLevel(currentVolume);
            player.start();//开始播放
        } catch (IOException e) {
            //TODO Auto-generated catch block
            e.printStackTrace();
        } catch (MediaException e) {
            //TODO Auto-generated catch block
            e.printStackTrace();
        }
    }
    /**
     * 为播放器自定义一个事件监听器
     */
    class PlayMusicListener implements PlayerListener {
        public void playerUpdate(Player player, String event, Object arg2) {
            try {
                if(event.equals(VOLUME_CHANGED)) {
                    //捕获到的音量改变事件
                    System.out.println("Player State Update : VOLUME_CHANGED");
                } else if(event.equals(STARTED)) {
                    //捕获到的开始播放事件
                    System.out.println("Player State Update : STARTED");
                } else if(event.equals(STOPPED)) {
                    //捕获到的停止播放事件
                    System.out.println("Player State Update : STOPPED");
                }
            } catch (Exception e) {
                e.printStackTrace();
            }
        }
    }
}
```

将文件名为 1.mid 的音频文件放在资源文件夹中，程序启动后，界面上显示一个由 Gauge 实现的音量控制界面。音乐播放画面展示如图 9-3 所示。当选择菜单 Play 后，程序则调用 playMusic()方法播放音乐。在该方法中通过 Manager 类调用静态方法，从一个数据流中获得播放器对象 player；然后给 player 对象添加事件监听器和音量控制器 volumeControl。完成实例化后启动，音乐开始播放。

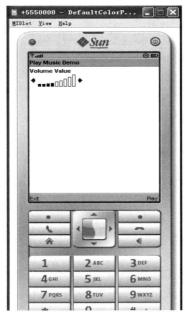

图 9-3　音乐播放画面展示

在播放过程中，我们可以通过左右键控制音量。程序中有一个单独的线程每隔 100ms 就检查一次音量是否改变。如果改变就会利用音量控制器 volumeControl 调整音量的大小。此外我们还可以利用菜单命令暂停或继续播放音乐。

注册了的事件监听接口可以捕获音量改变事件 VOLUME_CHANGED、播放事件 STARTED 和暂停事件 STOPED，并输出到控制台上。

9.3　视频播放

与音频的处理类似，视频处理由 Manager 指定视频数据来源，创建视频播放器 Player，同时获得自身的控制器 VideoControl，实现在高层界面或低层界面中显示视频。

典型的代码框架如下：

```
try {
    Player videoPlayer = Manager.createPlayer("videoDataLocation");
```

```
            //由数据源位置创建播放器
            videoPlayer.realize();
            //将播放器实例化
            VideoControl c = (VideoControl)videoPlayer.getControl("videoControl");
            //获得播放器自身的视频控制接口
            if(c != null) {
            c.initDisplayMode(VideoControl.USE_DIRECT_VIDEO, myCanvas);
            //初始化播放模式
            c.setVisible(true);                //视频内容可见
            videoPlayer.start();               //启动播放器
            }
    }
    catch(IOException ioe) {}              //捕获 I/O 异常
    catch(MediaException me) {}            //捕获媒体异常
```

与音频播放不同的是，视频播放在创建播放器以后需要获得一个视频控制接口，该接口用于控制视频的播放。在播放前，需要调用该接口的 initDisplayMode()方法指定视频播放的模式和调用 setVisible()方法使视频内容可见。

指定视频播放模式的方法：

Public Object initDisplayMode(int mode , Object arg);

参数 mode 指定显示模式，视频播放模式有两种，一种是在资源受限的设备上的播放模式，使用静态常量 USE_DIRECT_VIDEO 表示，该常量支持在 Canvas 画布上添加视频组件，这时需要第二个参数传递一个 Canvas 实例，指定容纳视频的画布。另一种是在非资源受限的设备上的播放模式，使用静态常量 USE_GUI_PRIMITIVE 表示，用于返回 GUI 界面组件。这时第二个参数可以设为 null。

设置视频可视的方法：

Public void setVisible(boolean b);

在默认的情况下，视频内容在画布上是不可见的，使用此方法可以显示/隐藏视频内容，参数是布尔型变量，为 true 时表示显示，为 false 时表示隐藏。

【示例程序 9-2】

下面的示例程序实现了一个手机在线视频播放器，能够根据 URL 的地址，连接服务器获取视频文件，并在手机上播放。实现视频的 MIDlet 类为：VideoMIDlet.java。

程序文件：

```
import java.io.InputStream;
import javax.microedition.io.Connector;
import javax.microedition.io.HttpConnection;
import javax.microedition.lcdui.*;
import javax.microedition.media.*;
import javax.microedition.media.control.*;
```

```java
import javax.microedition.midlet.*;

public class VideoMIDlet extends MIDlet implements CommandListener, PlayerListener, Runnable {
    private Display display;
    private Form form;
    private TextField url;
    private Command start = new Command("Play", Command.SCREEN, 1);
    private Command stop = new Command("Stop", Command.SCREEN, 2);
    private Player player;

    public VideoMIDlet() {
        display = Display.getDisplay(this);
        form = new Form("Demo Player");
        //创建输入视频 URL 地址的文本框
        url = new TextField("Enter URL:", "", 100, TextField.URL);
        form.append(url);
        form.addCommand(start);
        form.addCommand(stop);
        form.setCommandListener(this);
        display.setCurrent(form);
    }

    protected void startApp() {
        try {
            //判断如果播放器处于就绪状态，就播放视频
            if (player != null && player.getState() == Player.PREFETCHED) {
                player.start();
            } else {
                defplayer();
                display.setCurrent(form);
            }
        } catch (MediaException me) {
            reset();
        }
    }

    protected void pauseApp() {
        try {
            //如果播放器处于播放状态，则停止播放
            if (player != null && player.getState() == Player.STARTED) {
```

```java
                player.stop();
            } else {
                defplayer();
            }
        } catch (MediaException me) {
            reset();
        }
    }

    protected void destroyApp(boolean unconditional) {
        form = null;
        try {
            defplayer();
        } catch (MediaException me) {
        }
    }

    public void playerUpdate(Player player, String event, Object data) {
        if (event == PlayerListener.END_OF_MEDIA) {
            try {
                defplayer();
            } catch (MediaException me) {
            }
            reset();
        }
    }

    public void commandAction(Command c, Displayable d) {
        if (c == start) {
            start();
        } else if (c == stop) {
            stopPlayer();
        }
    }

    public void start() {
        //创建播放线程
        Thread t = new Thread(this);
        t.start();
    }
```

```java
//为了防止阻塞,网络连接应该被定义在一个线程中,而不是在 commandAction()方法中
public void run() {
    play(getURL());
}

String getURL() {
    return url.getString();
}

void play(String url) {
    try {
        VideoControl vc;
        defplayer();
        InputStream dis = null;
        //建立并打开 HTTP 连接
        HttpConnection con = (HttpConnection) Connector.open(url,Connector.READ);
        //打开网络输入流
        dis = con.openInputStream();
        if (dis != null) {
            //创建播放器实例
            player = javax.microedition.media.Manager.createPlayer(dis,"video/mpeg");
            //添加消息监听器
            player.addPlayerListener(this);
            //准备播放信息
            player.realize();
            //创建视频控制接口
            vc = (VideoControl) player.getControl("VideoControl");
            if (vc != null) {
                //得到 GUI 界面组件
                Item video = (Item) vc.initDisplayMode(VideoControl.USE_GUI_PRIMITIVE,null);
                Form v = new Form("Playing Video...");
                StringItem si = new StringItem("Status: ", "Playing...");
                v.append(si);
                //将媒体播放器组件添加到屏幕上
                v.append(video);
                display.setCurrent(v);
            }
        }
        player.prefetch();
```

```
        //播放视频
            player.start();
        } catch (Throwable t) {
            System.out.println(t);
            reset();
        }
    }

    void defplayer() throws MediaException {
        if (player != null) {
            if (player.getState() == Player.STARTED) {
                player.stop();
            }
            if (player.getState() == Player.PREFETCHED) {
                player.deallocate();
            }
            if (player.getState() == Player.REALIZED
                    || player.getState() == Player.UNREALIZED) {
                player.close();
            }
        }
        player = null;
    }
    void reset() {
        player = null;
    }

    void stopPlayer() {
        try {
            defplayer();
        } catch (MediaException me) {
        }
        reset();
    }
}
```

在运行之前先将一个 mpg 格式的视频文件 dlmu.mpg 放到 Tomcat 服务器的根目录下。本书设置的根目录在 D:\mobile 中，并设置 http 的本机接口为 8066。启动 Tomcat 服务器虚目录为 mobile，然后运行 VideoMIDlet.java。如图 9-4 所示，为获取视频文件 URL 地址输入 http://127.0.0.1:8066/mobile/dlmu.mpg，图 9-5 为视频文件播放效果。

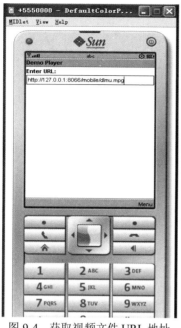

图 9-4　获取视频文件 URL 地址　　　　图 9-5　视频文件播放效果

9.4　手机拍照的实现

现在越来越多的手机都支持拍照功能,下面我们就来介绍如何使用 MMAPI 开发手机摄像头拍照功能。

第一步,创建 Player 实例:

Player mPlayere = Manager.createPlayer("capture://video");

参数"capture://video"是启动摄像头进行图像捕获的协议。

第二步,调用 Player 的 realize()方法,使摄像头处于就绪状态,具体代码:

mPlayer.realize();

最后,创建 VideoControl 视频控制器对象 vc,并通过它创建 GUI 视频组件,代码如下:

VideoControl vc=(VideoControl)mplayer.getControl("VideoControl");
If(vc!=null) {
vc.initDisplayMode(VideoControl.USE_DIRECT_VIDEO, this);
//创建 GUI 组件
vc.setDisplaySize(128, 160)
//设置拍照显示窗口大小
}
vc.setVisible(true); //设置可见性
mplayer.start();　　//可以启动拍照

摄像头启动之后，还要进一步定义捕获图像进行拍照的代码。

通过视频控制器对象调用 getSnapshot 方法，启动摄像头捕获图像，其方法为：

Public byte[] getSnapshot(String imageType)throws MediaException

参数 imageType 为照片的格式，如果为 null 则采用手机默认格式（一般为 PNG 格式），该方法返回图片的二进制数据。如上述代码控制器对象为 vc，则可以用下面的代码获得照片的二进制数据。

Byte[] raw = vc.getSnapshot(null);

将摄像头获取的二进制数组转换为 Image 类对象：

Image image = Image.createImage(raw, 0, raw.length);

在开发照相机程序时要搞清楚目标机型是否支持照相功能，如果支持，那么支持的图像格式及图像的大小是多少。下面的代码就来解决这个问题。如果 prop2 的值不为空，则说明手机支持照相功能，prop2 的值就是照相机支持的图像格式。

```
String prop2 = System.getProperty("video.snapshot.encodings");
if (prop2!=null) {
    System.out.println("照相机支持的图像格式：" + prop2);
}
```

以上代码在 WTK 缺省的彩色模拟器上的输出结果为：

照相机支持的图像格式：

encoding=pcm

encoding=pcm&rate=8000&channels=1

encoding=pcm&rate=22050&bits=16&channels=2

在照相机拍摄图像时，可以通过 setDisplaySize 函数设置捕获图像的大小，其语法定义如下：

Public void setDisplaySize(int width, int height)throws MediaException

其中参数 width 和 height 为捕获图像的大小。

【示例程序 9-3】

下面给出一个手机拍照的示例程序，Snapper 由两个文件组成：第一个是画布类程序：CameraCanvas.java；第二个是 MIDlet 类：SnapperMIDlet.java。

画布类 CameraCanvas.java：

```java
import javax.microedition.lcdui.*;
import javax.microedition.media.MediaException;
import javax.microedition.media.control.VideoControl;

public class CameraCanvas extends Canvas {
    private SnapperMIDlet mSnapperMIDlet;

    public CameraCanvas(SnapperMIDlet midlet, VideoControl videoControl) {
        int width = getWidth();
        int height = getHeight();
```

```
        mSnapperMIDlet = midlet;

        videoControl.initDisplayMode(VideoControl.USE_DIRECT_VIDEO, this);
        try {
            videoControl.setDisplayLocation(2, 2);
            videoControl.setDisplaySize(width - 4, height - 4);
        }
        catch (MediaException me) {
            try { videoControl.setDisplayFullScreen(true); }
            catch (MediaException me2) {}
        }
        videoControl.setVisible(true);
    }

    public void paint(Graphics g) {
        int width = getWidth();
        int height = getHeight();

        //绘制 VideoControl 控件的边框
        g.setColor(0x00ff00);
        g.drawRect(0, 0, width - 1, height - 1);
        g.drawRect(1, 1, width - 3, height - 3);
    }

    public void keyPressed(int keyCode) {
        int action = getGameAction(keyCode);
        if (action == FIRE)
            mSnapperMIDlet.capture();
    }
}
```

MIDlet 类 SnapperMIDlet.java：

```
import javax.microedition.lcdui.*;
import javax.microedition.media.*;
import javax.microedition.media.control.*;
import javax.microedition.midlet.MIDlet;

public class SnapperMIDlet extends MIDlet implements CommandListener,Runnable {
    private Display mDisplay;
    private Form mMainForm;
```

```java
private Command mExitCommand, mCameraCommand;
private Command mBackCommand, mCaptureCommand;
//创建播放器对象
private Player mPlayer;
//创建视频控制器接口
private VideoControl mVideoControl;

public SnapperMIDlet() {
    mExitCommand = new Command("Exit", Command.EXIT, 0);
    mCameraCommand = new Command("Camera", Command.SCREEN, 0);
    mBackCommand = new Command("Back", Command.BACK, 0);
    mCaptureCommand = new Command("Capture", Command.SCREEN, 0);

    mMainForm = new Form("Snapper");
    mMainForm.addCommand(mExitCommand);
    String supports = System.getProperty("video.snapshot.encodings");
    if (supports != null && supports.length() > 0) {
        mMainForm.append("Ready to take pictures.");
        mMainForm.addCommand(mCameraCommand);
    }
    else
        mMainForm.append("Snapper cannot use this " + "device to take pictures.");
    mMainForm.setCommandListener(this);
}

public void startApp() {
    mDisplay = Display.getDisplay(this);
    mDisplay.setCurrent(mMainForm);
}

public void pauseApp() {}

public void destroyApp(boolean unconditional) {  }

public void commandAction(Command c, Displayable s) {
    if (c.getCommandType() == Command.EXIT) {
        destroyApp(true);
        notifyDestroyed();
    }
```

```java
        else if (c == mCameraCommand)
            showCamera();          //启动摄像头
        else if (c == mBackCommand)
            mDisplay.setCurrent(mMainForm);    //显示主屏幕界面
        else if (c == mCaptureCommand) {
            new Thread(this).start();   //创建摄像头捕获图像的线程
        }
    }

    private void showCamera() {
        try {
            //创建播放器对象
            mPlayer = Manager.createPlayer("capture://video");
            mPlayer.realize();     //使摄像头处于就绪状态
            //创建视频控制器接口
            mVideoControl = (VideoControl)mPlayer.getControl("VideoControl");
            Canvas canvas = new CameraCanvas(this, mVideoControl); //新建 Canvas 画布对象
            canvas.addCommand(mBackCommand);
            canvas.addCommand(mCaptureCommand);
            canvas.setCommandListener(this);
            mDisplay.setCurrent(canvas);
            mPlayer.start();//启动摄像头
        }
        catch (Exception ioe) { handleException(ioe); }
        //catch (MediaException me) { handleException(me); }
    }
    //捕获图像的线程中的方法定义
    public void run(){
        capture();
    }
    public void capture() {
        try {
            //得到摄像头拍照的图像的字节数组
            byte[] raw = mVideoControl.getSnapshot(null);
            //将摄像头获取的图像二进制字节数组转换成 Image 对象
            Image image = Image.createImage(raw, 0, raw.length);
            Image thumb = createThumbnail(image);   //调用图像转换方法

            //将图像显示在屏幕上
            if (mMainForm.size() > 0 && mMainForm.get(0) instanceof StringItem)
```

```
            mMainForm.delete(0);
            mMainForm.append(thumb);
        mDisplay.setCurrent(mMainForm);
        mPlayer.close();    //关闭播放器
        mPlayer = null;
        mVideoControl = null;
    }
    catch (MediaException me) { handleException(me); }
}

private void handleException(Exception e) {
    Alert a = new Alert("Exception", e.toString(), null, null);
    a.setTimeout(Alert.FOREVER);
    mDisplay.setCurrent(a, mMainForm);
}
//图像转换方法
private Image createThumbnail(Image image) {
    int sourceWidth = image.getWidth();
    int sourceHeight = image.getHeight();
    int thumbWidth = 64;
    int thumbHeight = -1;
    if (thumbHeight == -1)
        thumbHeight = thumbWidth * sourceHeight / sourceWidth;
    Image thumb = Image.createImage(thumbWidth, thumbHeight);
    Graphics g = thumb.getGraphics();

    for (int y = 0; y < thumbHeight; y++) {
        for (int x = 0; x < thumbWidth; x++) {
            g.setClip(x, y, 1, 1);
            int dx = x * sourceWidth / thumbWidth;
            int dy = y * sourceHeight / thumbHeight;
            g.drawImage(image, x - dx, y - dy, Graphics.LEFT | Graphics.TOP);
        }
    }
    Image immutableThumb = Image.createImage(thumb);
    return immutableThumb;
}
```

启动手机拍照主界面如图9-6所示，摄像头准备就绪界面如图9-7所示，显示手机拍摄图像如图9-8所示。

图 9-6 启动手机拍照主界面

图 9-7 摄像头准备就绪界面

图 9-8 显示手机拍摄图像

9.5 本章小结

本章主要分为 4 个部分：移动媒体 MMAPI 概述、音频的处理、视频的处理和手机拍照功能的实现。在 MMAPI 中，我们介绍了移动媒体程序设计的层次结构以及查询设备支持媒体的能力；接下来我们介绍了音频和视频的处理，实现了手机拍照的功能，并以实例演示了具体的构建方法，明确了管理器、播放器和控制器的概念和相互关系。

10 无线消息程序设计

本章主要介绍 Java ME 无线消息方面的内容。读者需要掌握以下知识点：
- 无线消息传递的基本原理
- WMA 的基本内容
- 使用无线消息 API 的基本方法
- WTK 提供的 WMA 测试工具
- SMS 消息的接收与发送
- CBS 消息的接收

10.1 无线消息概述

无线消息支持的 Java ME 应用程序能独立于平台访问无线资源，如使用全球移动通信系统 GSM（Global System for Mobile Communication）或码分多址 CDMA（Code Division Multiple Access）等无线网络资源，实现消息传递服务。这些消息大致可以分为两类：一对一的短信传递服务（Short Message Service，SMS）和在某个地区提供一对多消息传递的小区广播服务（Cell Broadcast Service，CBS）。

10.1.1 GSM 短消息服务

短消息服务是移动设备之间通过无线网络收发文本的一种服务，通常情况下，文本的长度不超过 160 个字符。GSM 短信的收发过程如图 10-1 所示。

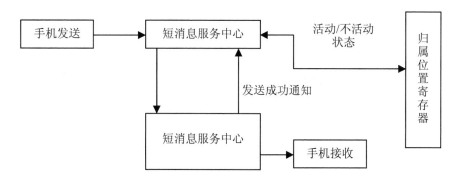

图 10-1　短消息发送和接收机

消息一经发送，就会被短消息服务中心接收。该中心根据发送目的地，获得目的地所属的位置和在线状态，这个过程通过查询"归属位置寄存器"完成。如果接收方处于离线状态，则服务中心保存此短信一段时间（一般是 24 小时），等到接收方在线时，再发送该消息，如果接收方处于在线状态，则直接执行发送过程。

发送过程是由"GSM 无线消息收发系统"（Message Delivery System）完成的，首先寻呼接收方设备，得到响应后发送消息，并向短消息服务中心返回"已发送"的信息。并将不再尝试发送第二次。从消息发送接收机制中可以看到：发送消息实际上是从手机端发向短消息服务中心。接收消息实际上是从 GSM 消息发送系统接收消息。因此我们在编程中模拟发送和接收只需要模拟器将短消息发送到 WTK 的控制台，和从控制台接收消息即可。

10.1.2　GSM 小区广播

GSM 小区广播负责将消息发送到某个区域内的所有接收方，这里提到的接收方也被称为移动站（Mobile Station）。GSM 的小区广播消息的格式可以是文本类型，也可以是二进制类型，最多发送 15 页数据，每页最多有 93 个字符。每条消息都包含以下 4 种属性：

- 频道号：表示消息主题的头部，例如"天气预报"或"交通信息"等。
- 消息代码：标识特定的消息，是一条消息的唯一标识。
- 更新号：确定消息的版本，在一些情况下，某些消息需要时时更新，版本号可以确认将最新的消息内容发送到接收端。
- 语言：指定消息所使用的语言。

了解短消息和小区广播的基础知识后，我们可以进一步介绍 WMA（Wireless Messaging API）无线消息 API。

10.2 WMA 概述

Wireless Messaging API（WMA）是属于 Java ME 的一个可选包（Optional Package），提供了无线通信的高级抽象，它隐蔽了传输层，因而编程所要做的工作只是创建消息、发送消息和接收消息。WMA 早期的 1.1 版（JSR120）仅支持文本短信和二进制短消息。而目前常用的 JSR205 中定义的 WMA 2.0 则增加了对发送和接收多媒体信息（Multi-media Message Service，MMS）的支持。

WMA 类库

WMA 2.0 的短消息开发包为 javax.wireless.messaging，定义了所有用于发送和接收无线消息的接口和类。以下是对主要接口和类的介绍：

（1）Message 接口

它定义了不同类型消息的基础，并派生了三个子接口：

- BinaryMessage 接口：是带有二进制有效载荷属性的消息对象。
- MultipartMessage 接口：表示一种多媒体通信的接口。我们可以随时添加和移出参加多方通信的一方。它是包含一个消息头和多个 MIME（Multipurpose Internet Mail Extensions）格式的消息体对象。
- TextMessage 接口：是带有文本有效载荷属性的消息对象。

此外，Message 接口还定义了一些基本的方法：

- getAddress()方法：获得消息的地址，对于发送消息，返回目的地址；对于接收消息，返回源地址。
- setAddress()方法：设置消息的地址（由参数指定），专指发送消息的目的地址。
- getTimestamp()方法：获得消息发送的时间，返回值是一个 Date 类型的变量。

实现此接口的对象实例表示了一种抽象的消息，采用转发的方式由发送方传递到接收方，并且从发送方能够获得消息发送的状态。

（2）MessageListener 接口

消息监听器提供了一种在接收到消息时自动产生通知的机制可实现异步接收消息，当某个消息到来时，自动激活 notifyIncomingMessage()方法。

（3）MessageConnection 接口

该接口继承自 javax.microedition.io.Connection，定义了消息发送和接收的基本方法。可以通过调用 Connector 的 open(URL)方法得到 MessageConnection 实例，调用 close()方法关闭连接。需要注意的是消息连接分为服务器端和客户端两类，它们之间通过 URL 进行区分，典型的客户端连接代码如下：

```
String url = "sms://138xx038620:50";
MessageConnection mcClient = (MessageConnection)Connector.open(url);
…
mcClient.close();
```

其中，sms 表示协议类型，138xx038620 表示客户端地址（电话号码），50 表示端口号。典型的服务器端连接代码如下：

```
String url = "sms://:50";
MessageConnection mcServer = (MessageConnection)Connector.open(url);
mcServer.close();
```

作为服务器端的参数 URL，只有协议类型和端口号，无须指定接收地址号码。

（4）MessagePart 类

该类是在 WMA 2.0 版本中被添加的，用于表示"多方短信通信的一方"（实际是多媒体消息中的一部分），其构造方法中的字节数组可以传输多媒体。其实例可以添加到 MultipartMessage 接口中，每一个实例包含内容、MIME 类型和 ID，还可以选择包含资源位置和编码模式等。

（5）SizeExceededException 类

当发送的消息超过容量时抛出的异常。

10.3　使用 WTK 中的 WMA 控制台

WTK 开发包提供了一些用于无线消息开发的工具，即 WMA 控制台，可以帮助我们开发和测试无线消息处理程序。首先介绍如何使用这些开发工具。

10.3.1　配置和启动 WTK 中的 WMA 控制台

在建立无线消息开发项目之前，需要对项目本身所包含的 MIDP 可选包进行配置，配置方法如下：在 Windows 的"开始"菜单中选择"程序"和"Wireless Toolkit 2.5.2_01 for CLDC"，打开"Preferences"，在左侧树形结构中选中"WMA"选项，如图 10-2 所示。

在"First Assigned Phone Number"文本框中列出了一个默认的电话号码，它代表了 WMA 控制台的电话号码。如需要还可以设置"Phone Number of Next Emulator"（下一个模拟器所代表的号码），允许为空。中间的滑块可以设置随机丢失信息的百分比，最后可以设置模拟信息发送的延迟时间，最后点击"OK"按钮保存设置。

要启动 WTK 中的 WMA 控制台可以在 Windows 的"开始"菜单中选择"程序"和"Wireless Toolkit 2.5.2_01 for CLDC"，打开"Utilities"将显示"Utilities"对话框，如图 10-3 所示。

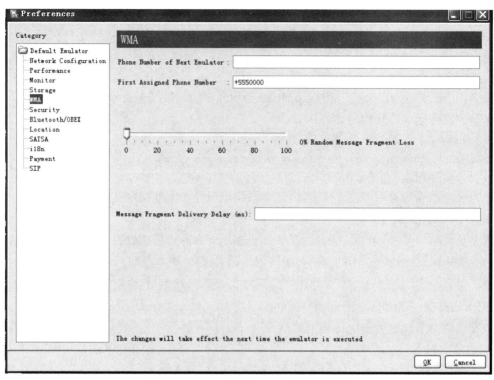

图 10-2　WTK 中设置 WMA 参数的界面

图 10-3　Utilities 对话框

双击 WMA Console（WMA 控制台）选项，打开 WMA Console 对话框如图 10-4 所示。

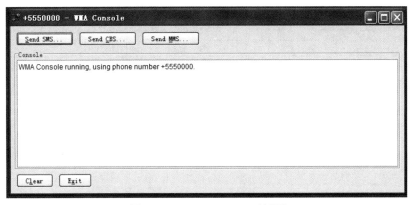

图 10-4　WMA Console 对话框

启动控制台后在文本框中默认显示 WMA 控制台已经运行及当前使用的模拟电话号码，可以"清除"其内容。在控制台上有三个按钮，分别用于发送文本消息、发送小区广播和发送多媒体消息。

10.3.2　使用 WMA 控制台发送文本消息

使用 WMA 控制台可以向手机发送文本消息，这个功能用于测试文本消息的发送和接收功能。点击图 10-4 中的"Send SMS"按钮，打开如图 10-5 所示的对话框，在"Text SMS"选项卡中，可以指定发送到的接收方号码、端口号和消息内容，点击"Send"按钮完成发送。

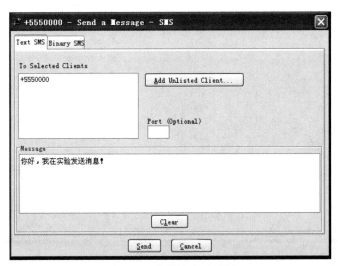

图 10-5　发送消息对话框

发送消息后在控制台中可以看到发送成功的提示,如图 10-6 所示。

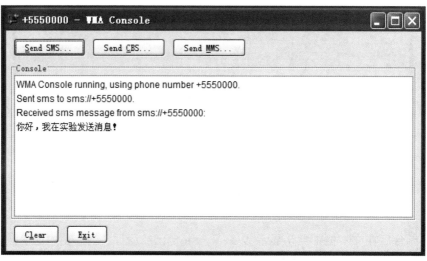

图 10-6　发送文本消息成功提示

发送二进制消息和发送文本消息在同一对话框中完成。只需点击"Binary SMS"选项卡,出现发送二进制消息的对话框,如图 10-7 所示。

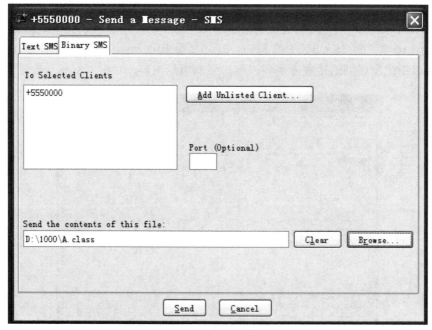

图 10-7　发送二进制消息对话框

发送二进制消息与发送文本消息相类似，只是发送文件内容需要输入二进制文件名，可以单击"Browse"通过"Open"选择要发送的二进制文件。选择后点击"Send"完成二进制消息发送，之后可以在控制台看到如图 10-8 所示发送二进制消息成功的提示。

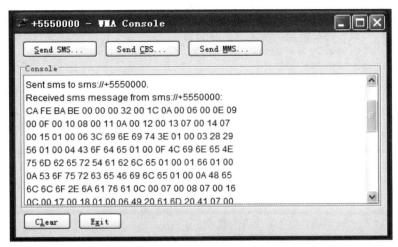

图 10-8　发送二进制消息成功提示

10.3.3　使用 WMA 控制台发送小区广播

在控制台中点击"Send CBS"按钮打开发送小区广播对话框，如图 10-9 所示。

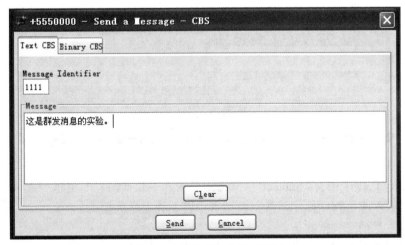

图 10-9　发送小区广播对话框

在"Message Identifier"文本框中输入消息标识号，在"Message"文本区中输入要广播的消息后点击"Send"按钮完成发送，控制台显示如图 10-10 所示发送小区广播成功对话框。

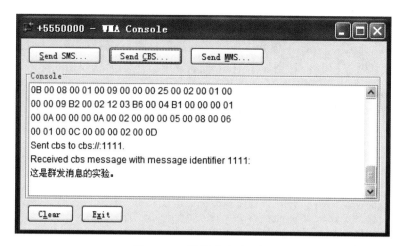

图 10-10　发送成功对话框

10.3.4　使用 WMA 控制台发送多媒体消息

在控制台中点击"Send MMS"按钮打开发送多媒体消息对话框，如图 10-11 所示。

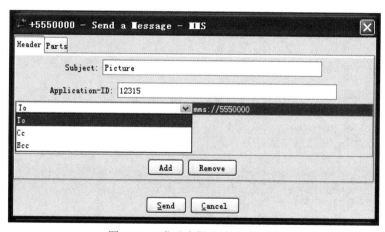

图 10-11　发送多媒体消息对话框

在"Header"选项卡中有"Subject"文本框，可输入多媒体消息的主题；在"Application-ID"文本框中可以输入应用程序标识，其下拉列表中有 3 种发送类型：

To：直接发送消息到接收方。

Cc：在发送消息的同时，把消息抄送到另外的目标地址，并且接收方可以看到抄送地址。

Bcc：密件抄送，与 Cc 相同，只是接收方看不到抄送地址。

在 mms:// 后面输入具体接收地址，并且必须以"mms://"开头，例如我们可以选择服务器

5550000 作为目标地址。点击"Add"按钮或"Remove"按钮可以添加或删除多个目标地址。

点击"Parts"选项卡，可以添加发送多媒体的内容，如图 10-12 所示。

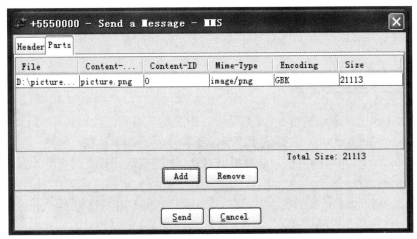

图 10-12　发送多媒体内容的"Parts"选项卡

点击"Add"按钮，可以添加要发送的多媒体文件，我们选取 D:\picture.png 图片文件。最后，点击"Send"按钮完成发送，在控制台可以看到，发送成功的有关信息如图 10-13 所示。在这些信息中有：消息的主题、优先级、多媒体部分编号、内容标识、MIME 类型、编码种类、消息长度等信息。

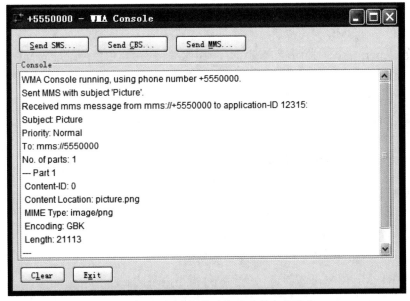

图 10-13　发送多媒体消息成功的有关信息

10.4 编写利用 WMA 控制台收发短消息的程序

10.3 节我们讲述了 WTK 中 WMA 控制台的配置，本节将讲述利用 WMA 2.0 编写各种类型消息发送和接收程序。

10.4.1 发送和接收 SMS 消息

使用 MIDlet 程序发送和接收 SMS 消息，首先要建立消息连接。在 WMA 2.0 中定义了消息连接类 MessageConnection，它提供了消息发送和接收的基本功能。包括：发送和接收的基本方法、创建 Message 对象的方法，以及根据 GSM 连接规范计算发送指定 Message 对象所需底层协议段数量的方法等。

构造 MessageConnection 类的实例时，要传递一定格式的 URL 字符串给 Connector.open(URL) 工厂方法。这个 URL 有三种格式：

（1）sms://+phone_number 格式：向特定手机号码发送消息。由于没有指定端口，消息会发送到默认端口，被手机中的消息应用程序接收。

注意：由于手机本地的消息应用程序拥有比 Java ME 程序更高的优先级，因此使用默认端口发送出的短消息不可能被 Java ME 应用程序监听到。

（2）sms://+phone_number:port 格式：向特定手机号码的特定端口号发送消息。如果收件人手机正在运行一个 Java ME 程序监听该端口，则该手机可以接收到该消息。

注意：在这种情况下，本地的 SMS 收件箱无法接收到这种特定端口的短消息，如果一部手机上需要运行多个 SMS 相关应用程序，一定要预防端口地址冲突。

（3）sms://:port 格式：监听特定端口。使用这种格式的 URL 时，Java ME 程序作为该特定端口的一个服务器，因此这个连接也成为服务器模式的连接。使用该连接的应用程序可以接收和发送给这个端口的 SMS 消息。

消息连接对象建立后，可以通过 MessageConnection 类的 newMessage()方法创建一个 Message 类的实例。该方法定义如下：

```
public Message newMessage(String type);
public Message newMessage(String type, String address);
```

其中，参数 address 指定消息接收方地址；参数 type 指定要创建 Message 消息的类型，可以有以下三个取值：

TEXT_MESSAGE：创建一个文本类型消息。

BINARY_MESSAGE：创建一个二进制类型消息。

MULTIPART_MESSAGE：创建一个多媒体类型消息。

Message 类型是个接口，在这个接口中定义了三个方法：

public String getAddress()：返回消息中的地址。

public void setAddress(String address)：设置消息接收地址。

public Date getTimestamp()：返回消息被发送的时间。

一个文本消息由三部分组成，如图 10-14 所示，消息的源/目标地址（发送时该字段保留源地址，接收时该字段保留目标地址），Payload 字段是消息的正文内容，或称为消息的有效载荷，最后是控制标志，表明该消息是否被阻塞。

源/目标地址	Payload	控制标志

图 10-14 文本消息的组成结构

TextMessage 接口作为 Message 接口的子接口，当然继承了其父接口的上述三个方法，并扩展了两个方法：

public String getPayloadText()：获取消息文本。

public void setPayloadText(String data)：设置消息文本。

设置消息后就可以使用 MessageConnection 的 send()方法将消息发送出去。不过在实际编写发送 SMS 消息时，一定将发送消息的代码放在主线程之外的一个新线程中，否则将会引起网络阻塞，使程序处于瘫痪，不能执行。

【示例程序 10-1】

下面通过一个完整的示例程序介绍如何发送和接收文本消息。

首先创建一个 TextMessageDemo 项目，并且 MIDlet 类名也为 TextMessageDemo.java。

```java
import javax.wireless.messaging.*;
import javax.microedition.io.*;
import java.io.*;
import java.lang.*;
import java.util.*;

public class TextMessageDemo extends MIDlet
        implements CommandListener, MessageListener,Runnable {
    private Command cmdExit;    /** 退出命令按钮 */
    private Command cmdSendMsg; /** 发送消息的命令按钮 */
    private Display display;    //Display 管理
    private Form form;
    private TextField tfMsgText;    /** 消息内容 */
    private TextField tfPhoneNumber; /** 用于输入发送到的电话号码 */
    private MessageConnection recieveConn; /** 用于接收消息 */

    public TextMessageDemo() {
        display = Display.getDisplay(this);
```

```java
            cmdExit = new Command("退出程序", Command.EXIT, 1);
            cmdSendMsg = new Command("发送消息", Command.SCREEN, 2);
            tfMsgText = new TextField("请输入消息内容:", "", 255, TextField.ANY);
            tfPhoneNumber = new TextField("请输入接收号码:", "", 255,
                    TextField.PHONENUMBER);
        }

        /**
         * 开始运行 MIDlet
         */
        public void startApp() {
            try {
                recieveConn = (MessageConnection)Connector.open
                        ("sms://:5008");
                //Register the listener for inbound messages.
                recieveConn.setMessageListener(this);
            } catch (IOException ioe) {
                System.out.println("不能进行接收消息连接:" + ioe.toString());
            }
            form = new Form("接收/发送消息演示 - 接收端口为 5008");
            form.append(tfMsgText);
            form.append(tfPhoneNumber);
            form.addCommand(cmdExit);
            form.addCommand(cmdSendMsg);
            form.setCommandListener(this);
            display.setCurrent(form);
        }

        public void pauseApp() {   }

        public void destroyApp(boolean unconditional) {
            notifyDestroyed();
        }
        /**
         * 处理命令按钮事件
         */
        public void commandAction(Command c, Displayable s) {
            if (c == cmdExit) {
                destroyApp(false);
            } else if (c == cmdSendMsg) {
                //检查电话号码是否存在
```

```java
                String pn = tfPhoneNumber.getString();
                if (pn.equals("")) { //注意如果使用 pn==""会不起作用
                    Alert alert = new Alert("发送消息错误","请输入接收的电话号码", null, AlertType.ERROR);
                    alert.setTimeout(2000);
                    display.setCurrent(alert, form);
                    AlertType.ERROR.playSound(display);
                } else {
                    try {
                        new Thread(this).start();
                    } catch (Exception exc) {
                        exc.printStackTrace();
                    }
                }
            }
        }
    }

    public void run() {
        boolean result = true;
        try{
            String address = "sms://+" + tfPhoneNumber.getString();//地址
            //建立连接
            MessageConnection conn=(MessageConnection)Connector.open(address);
            //设置短信息类型为文本
            TextMessage msg = (TextMessage)conn.newMessage(
                    MessageConnection.TEXT_MESSAGE);
            msg.setAddress(address); //设置消息地址
            msg.setPayloadText(tfMsgText.getString());//设置信息内容
            conn.send(msg); //发送消息
        }catch(Exception e){
            result = false;
            System.out.println("发送短消息错误: " + e.toString());
        }
    }

    /**
     * 接收消息
     */
    public void notifyIncomingMessage(MessageConnection conn) {
        System.out.println("收到了一条新信息");
        Message msg = null;
```

```
//读取消息
try {
    msg = conn.receive();
} catch (Exception e) { //处理读取消息时的异常
    System.out.println("读取消息错误: " + e);
}

//处理读取的文本消息
if (msg instanceof TextMessage) {
    TextMessage tmsg = (TextMessage)msg;
    form.append("接收到一条消息,发送时间: " +
            tmsg.getTimestamp().toString() + "\n");
    form.append("消息发送方: " + tmsg.getAddress() + "\n");
    form.append("消息内容: " + tmsg.getPayloadText());
}
```

在这个示例程序中首先展示模拟器接收和发送界面,如图 10-15 所示。

图 10-15　模拟器接收和发送界面

然后启动 SMS 短消息发送控制台,界面如图 10-16 所示。输入发送信息"测试发送消息"和端口号 5008,发送客户端手机地址 5550000,之后点击"Send"按钮。

第 10 章 无线消息程序设计

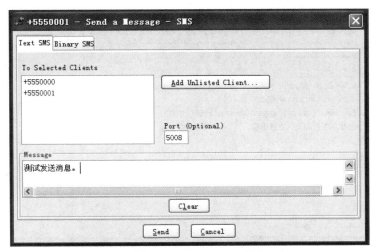

图 10-16 SMS 短消息发送控制台界面

模拟器显示收到消息，界面如图 10-17 所示。

在手机模拟器界面中输入要发送的消息"这是从客户端测试发送消息"和向服务器发送号码 5550001，之后点击"发送消息"按钮，模拟器发送界面如图 10-18 所示，服务器接收界面如图 10-19 所示。

图 10-17 模拟器显示收到消息

图 10-18 模拟器发送界面

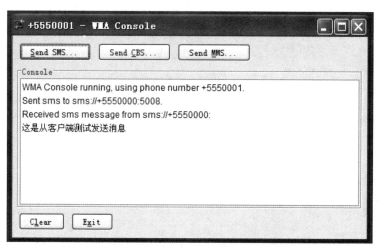

图 10-19　WMA 控制台服务器接收文本消息界面

10.4.2　发送和接收二进制消息

二进制消息在实际应用中具有非常重要的作用，可以传输许多文本消息所不能传输的信息，如传递二进制数据文件、图像文件和声音文件等。

二进制消息使用的是 BinaryMessage 接口，因此在接口设置和还原消息有效载荷内容方面与文本消息不同。

当 MessageConnection 类的实例调用 newMessage()方法，参数取值为 Message.Connection.BINARY_MESSAGE 时将得到 BinaryMessage 接口的实例。

BinaryMessage 接口除继承了 Message 接口中方法之外，还在 BinaryMessage 接口中新定义了两个方法：

public void setPayloadData(byte[] data);

public byte[] getPayloadData();

前一个方法将二进制消息内容放到一个字节数组，之后将这个字节数组作为二进制消息的有效载荷。在接收到二进制消息后使用后一个方法获得字节数组中的二进制消息的内容。

对于二进制消息通常要进行转换才能得到有效的数据，例如需要将收到的二进制数据转换为图像，然后才能显示到屏幕上。

【示例程序 10-2】

下面通过一个发送和接收二进制消息的示例程序展示二进制消息发送和接收的基本方法。项目名称为：BinMeg，MIDlet 类名为 BinMeg.java，源程序如下：

```
import javax.microedition.midlet.*;
import javax.microedition.lcdui.*;
import javax.microedition.io.*;
```

```java
import javax.wireless.messaging.*;
import java.io.*;

public class BinMsg extends MIDlet
        implements CommandListener, MessageListener,Runnable {
    private Command cmdExit;           /** 退出命令按钮 */
    private Command cmdSendMsg;        /** 发送消息的命令按钮 */
    private Display display; //Display 管理
    private Form form;
    private TextField tfMsgText; /** 消息内容 */
    private TextField tfPhoneNumber; /** 用于输入发送到的电话号码 */
    private MessageConnection recieveConn; /** 用于接收消息 */
    //在收到二进制消息时,以下一个表单和两个按钮用于显示提示信息
    Command cmdRead ;
    Command cmdReturn;
    Form    formNotice;
    BinMsgCanvas bmc = new BinMsgCanvas();//用于显示收到的图像
    private BinaryMessage bm; /** 保存接收到的二进制消息 */
    public BinMsg() {
        display = Display.getDisplay(this);
        cmdExit = new Command("退出程序", Command.EXIT, 1);
        cmdSendMsg = new Command("二进制格式发送", Command.SCREEN, 2);
        tfMsgText = new TextField("请输入消息内容：", "", 255, TextField.ANY);
        tfPhoneNumber = new TextField("请输入接收号码：", "", 255,TextField.PHONENUMBER);
    }
    /**
     * 开始运行 MIDlet
     */
    public void startApp() {
        //创建消息连接
        try {
            recieveConn = (MessageConnection)Connector.open
                    ("sms://:5008");
            //注册接收消息的监听器
            recieveConn.setMessageListener(this);
        } catch (IOException ioe) {
            System.out.println("不能进行接收消息连接：" + ioe.toString());
        }
        //初始化收到二进制消息时的提示窗口
```

```java
        cmdRead = new Command("阅读消息", Command.SCREEN, 2);
        cmdReturn = new Command("返回", Command.STOP, 1);
        formNotice = new Form("收到新二进制消息");
        formNotice.addCommand(cmdRead);
        formNotice.setCommandListener(this);

        //初始化显示图像的画布
        bmc.addCommand(cmdReturn);
        bmc.setCommandListener(this);
        //初始化主窗口
        form = new Form("接收/发送消息演示 - 接收端口为 5008");
        form.append(tfMsgText);
        form.append(tfPhoneNumber);
        form.addCommand(cmdExit);
        form.addCommand(cmdSendMsg);
        form.setCommandListener(this);
        display.setCurrent(form);
    }

    public void pauseApp() {  }

    public void destroyApp(boolean unconditional) {
        notifyDestroyed();
    }

    /**
     * 处理命令按钮事件
     */
    public void commandAction(Command c, Displayable s) {
        String label = c.getLabel();
        if (label.equals("退出程序")) {
            destroyApp(false);
        } else if (label.equals("二进制格式发送")) {
            //检查电话号码是否存在
            String pn = tfPhoneNumber.getString();
            if (pn.equals("")) { //注意如果使用 pn==""会不起作用
                Alert alert = new Alert("发送消息错误",
                        "请输入接收的电话号码", null,
                        AlertType.ERROR);
```

```java
                alert.setTimeout(2000);
                display.setCurrent(alert, form);
                AlertType.ERROR.playSound(display);
            } else {
                try {
                    new Thread(this).start();
                } catch (Exception exc) {
                    exc.printStackTrace();
                }
            }
        } else if (label.equals("阅读消息")) {    //显示二进制消息内容
            if (bm != null) {
                bmc.setPicData(bm.getPayloadData());
                display.setCurrent(bmc);
                bmc.repaint();
            }
        } else if (label.equals("返回")) { //返回主界面
            bm = null;
            display.setCurrent(form);
        }
    }
    /**
     * 定义线程的执行方法，给指定号码发送二进制信息
     */
    public void run() {
        try{
            String address = "sms://+" + tfPhoneNumber.getString();//地址
            //建立连接
            MessageConnection conn=(MessageConnection)Connector.open(address);
            //设置短信息类型为二进制类型
            BinaryMessage msg = (BinaryMessage)conn.newMessage(
                    MessageConnection.BINARY_MESSAGE);
            msg.setAddress(address); //设置消息地址
            msg.setPayloadData(tfMsgText.getString().getBytes());//设置信息内容
            conn.send(msg);     //发送消息
        }catch(Exception e){
            System.out.println("发送短消息错误：" + e.toString());
        }
    }
}
```

```java
/**
 * 接收消息
 */
public void notifyIncomingMessage(MessageConnection conn) {
    System.out.println("收到了一条新信息");
    Message msg = null;
    //读取消息
    try {
        msg = conn.receive();
    } catch (Exception e) {    //处理读取消息时的异常
        System.out.println("读取消息错误：" + e);
    }
    //处理读取的二进制消息
    if (msg instanceof BinaryMessage) {
        bm = (BinaryMessage)msg;
        formNotice.deleteAll();
        formNotice.append("发送时间：\n" +bm.getTimestamp().toString() + "\n");
        formNotice.append("消息发送方：\n" + bm.getAddress() + "\n");
        display.setCurrent(formNotice);
    }
}

class BinMsgCanvas extends Canvas {
    byte[] picData = null;
    //构造二进制消息显示画布，data 必须为二进制图形数据
    public BinMsgCanvas(byte[] data) {
        picData = data;
    }
    public BinMsgCanvas() {         }
    /**
     * 设置要显示的图形数据
     * 其中参数 data 为二进制图形数据
     */
    public void setPicData(byte[] data) {
        picData = data;
    }
    protected void paint(Graphics g) {
        if (picData != null) {
            ByteArrayInputStream bais=new ByteArrayInputStream(picData);
```

```
            Image img;
            try {
                img = Image.createImage(bais);
            } catch (Exception e) {
                System.out.println("读取图像数据错误: " + e.toString());
                return;
            }
            //显示图像
            g.drawImage(img,img.getWidth(), img.getHeight(),Graphics.LEFT | Graphics.TOP);
        }
    }
  }
}
```

示例程序初始运行界面如图 10-20 所示。

图 10-20　程序初始运行界面

在模拟器手机"请输入消息内容"文本框中输入"Hello World",在"请输入接收号码"中输入"5550001",点击"二进制格式发送"按钮,这时在 WMA 控制台上可以看到接收到的二进制信息,如图 10-21 所示。

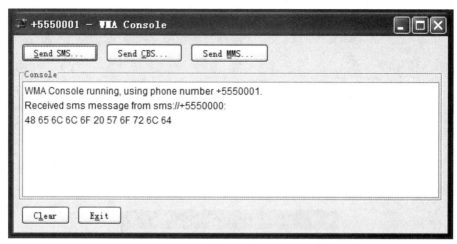

图 10-21　控制台接收到的二进制信息

从控制台发送一张图片的二进制信息:选取"Binary SMS"选项卡,如图 10-22 所示,填入端口号为"5008",选取一张存于 D 盘的图片文件"fire.png",点击"Send"按钮。

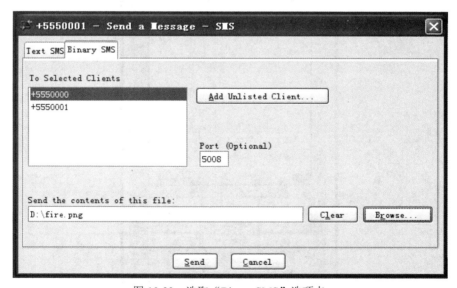

图 10-22　选取"Binary SMS"选项卡

在模拟器中收到二进制消息如图 10-23 所示。

点击"阅读消息"按钮,得到从控制台发送的图片消息,如图 10-24 所示。

图 10-23　模拟器中收到二进制消息　　　图 10-24　得到从控制台发送的图片消息

10.4.3　发送和接收多媒体消息

在 WMA 2.0 API 中新增加了 MUltipartMessage 类和 MessagePart 类，MUltipartMessage 类用来封装多媒体消息的操作和功能，MessagePart 类则封装了多媒体消息中的各个部分。

每一个多媒体消息都有一个消息头，还可以有一个或多个 MessagePart，并且每个 MessagePart 都由头和正文两部分组成。多媒体消息的结构如图 10-25 所示。

要发送和接收多媒体消息，首先创建多媒体消息的连接对象，这一点与短消息类似，仍然使用 Connector.open(URL)方法，不过创建多媒体消息连接的 URL 采用如下格式：

mms://+phone_number:appID

其中 mms:// 表示多媒体消息协议，phone_number 为接收端的号码，appID 为接收端的应用程序 ID 号，其最大长度为 32 个字符，一般为应用程序 java 类名。多媒体消息的应用程序 ID 与文本短消息的端口号等价。

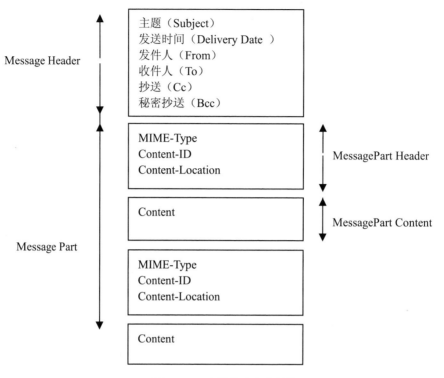

图 10-25 多媒体消息的结构

有了连接对象如 mmsconn 之后,可以使用 newMessage()方法,使用如下语句创建多媒体消息 MUltipartMessage 类对象:

MUltipartMessage mmmessage=(MUltipartMessage)
mmsconn.newMessage(MessageConnection.MULTIPART_MESSAGE);

MUltipartMessage 类中的 setAddress()方法和 addAddress()方法用于设置接收方地址,其具体定义如下:

public Boolean addAddress(String type, String address);
public void setAddress(String address);

参数 address 表示接收方地址,参数 type 表示地址类型,类型为"to"表示直接发送到,为"cc"表示发送的同时抄送到,为"bcc"表示加密抄送到。

设置好地址后,可以使用下面两个方法设置和返回主题:

public void setSubject(String subject);
public String getSubject();

MUltipartMessage 类中的 getAddress()和 getAddresses()可以返回发送方的地址,其方法定义如下:

public String getAddress();

```
public String[] getAddresses(String type);
```
如果 MUltipartMessage 类对象是收到的消息，则第一个方法返回的是发送方的地址，如果是发送的消息，第一个方法返回的是接收方的地址；第二个方法可以返回指定类型（to 类型、cc 类型或 bcc 类型）的地址。

下面的三个方法，可以删除设置的地址：
```
public Boolean removeAddress(String type, String address);
public void removeAddresses();
public void removeAddresses(String type);
```
第一个方法删除指定类型的指定地址；第二个方法没有参数，将删除全部地址；第三个方法将删除指定类型的地址。

向 MUltipartMessage 类对象中添加和删除 MessagePart 对象的方法，定义如下：
```
public void addMessagePart(MessagePart part );
public Boolean removeMessagePart(MessagePart part);
```
参数 part 是 MessagePart 类的实例。

MessagePart 对象共有 3 个构造方法：
```
public MessagePart(InputStream is, String mimeType, String contentID, String contentLocation, String enc )throws java.io.IOException,SizeExceededException;
public MessagePart(byte[] contents, String mimeType, String contentID, String contentLocation, String enc )throws SizeExceededException;
public MessagePart(byte[] contents, int offset, int lrngth, String mimeType, String contentID, String contentLocation, String enc) throws SizeExceededException;
```
构造方法中各参数含义介绍如下：

参数 is 是一个输入流对象，它以流的形式保存内容数据。

参数 mimeType 定义了内容的 MIME 类型，以保证接收方能够正确识别内容的格式，而不需要像二进制消息那样，接收端要预先知道内容数据的含义。常用的 MIME 类型有：

image/bmp：表示 BMP 图片

image/jpg：表示 JPEG 图片

image/gif：表示 GIF 图片

audio/x-wav：表示 WAV 音频格式

audio/midi：表示 MIDI 音频格式

audio/mp3：表示 MP3 音频格式

video/mpeg：表示 MPEG 视频格式

参数 contentID 指定头部的 content-id 域的值。

参数 contentLocation 指定附件文件的文件名和路径，如果该参数为 null，则 MessagePart 不会附加任何文件。

参数 enc 定义使用的编码方式，如 UTF-8 等。

参数 contents 是一个字节数组，其内容可以是图片、音频、视频等多媒体数据。

参数 offset 指定读取 contents 中多媒体数据的起始位置。

参数 length 定义了从 contents 中读取多媒体数据的长度。

【示例程序 10-3】

下面我们给出一个接收多媒体消息的示例程序，项目名：MMSReceive，应用程序名：MMSReceive.java。源程序如下：

```java
import javax.microedition.midlet.*;
import javax.microedition.io.*;
import javax.microedition.lcdui.*;
import javax.wireless.messaging.*;
import java.io.IOException;
public class MMSReceive extends MIDlet
    implements CommandListener, Runnable, MessageListener {

    private static final Command CMD_EXIT  =new Command("Exit", Command.EXIT, 2);
    private Form content;
    private Display display;
    private Thread thread; //声明接收多媒体消息的线程
    private String[] connections;
    private boolean done; //声明接收消息成功的标识位
    private String appID; //声明监听多媒体消息的 applicationID
    private MessageConnection mmsconn; //声明多媒体消息连接对象
    private Message msg; //声明获得的多媒体消息的对象
    private String senderAddress; //声明多媒体消息发送方的地址
    private Alert sendingMessageAlert;
    private Displayable resumeScreen;
    private String subject; //声明接收多媒体消息的标题
    private String contents; //声明接收多媒体消息的内容

    public MMSReceive() {
        appID = "MMSTest";   //初始化多媒体消息的 applicationID
        display = Display.getDisplay(this);
        content = new Form("MMS Receive");
        content.addCommand(CMD_EXIT);
        content.setCommandListener(this);
        content.append("Receiving...");
        sendingMessageAlert = new Alert("MMS", null, null, AlertType.INFO);
        sendingMessageAlert.setTimeout(5000);
        sendingMessageAlert.setCommandListener(this);
        resumeScreen = content;
    }
```

```java
public void startApp() {
    String mmsConnection = "mms://:" + appID;    //多媒体消息连接字符串
    if (mmsconn == null) {    //打开连接
        try {
            mmsconn = (MessageConnection) Connector.open(mmsConnection);
            mmsconn.setMessageListener(this);
        } catch (IOException ioe) {
            ioe.printStackTrace();
        }
    }
    //等待发送方发送多媒体消息
    connections = PushRegistry.listConnections(true);
    if (connections == null || connections.length == 0) {
        content.deleteAll();
        content.append("Waiting for MMS on applicationID " + appID + "...");
    }
    done = false;
    thread = new Thread(this);
    thread.start();
    display.setCurrent(resumeScreen);
}
/**  当多媒体消息到来时，触发该监听方法   */
public void notifyIncomingMessage(MessageConnection conn) {
    if (thread == null && !done) {
        thread = new Thread(this);
        thread.start();
    }
}
//读取多媒体消息线程的执行方法
public void run() {
    try {
        msg = mmsconn.receive();
        if (msg != null) {
            senderAddress = msg.getAddress();
            content.deleteAll();
            String titleStr = senderAddress.substring(6);
            int colonPos = titleStr.indexOf(":");
            if (colonPos != -1) {
                titleStr = titleStr.substring(0, colonPos);
            }
            content.setTitle("From: " + titleStr);
            if (msg instanceof MultipartMessage) {
```

```java
                        MultipartMessage mpm = (MultipartMessage)msg;
                        StringBuffer buff = new StringBuffer("Subject: ");
                        buff.append((subject = mpm.getSubject()));
                        buff.append("\nDate: ");
                        buff.append(mpm.getTimestamp().toString());
                        buff.append("\nContent:");
                        StringItem messageItem = new StringItem("Message",buff.toString());
                        messageItem.setLayout(Item.LAYOUT_NEWLINE_AFTER);
                        content.append(messageItem);
                        MessagePart[] parts = mpm.getMessageParts();
                        if (parts != null) {
                            for (int i = 0; i < parts.length; i++) {
                                buff = new StringBuffer();
                                MessagePart mp = parts[i];
                                buff.append("Content-Type: ");
                                String mimeType = mp.getMIMEType();
                                buff.append(mimeType);
                                String contentLocation = mp.getContentLocation();
                                buff.append("\nContent:\n");
                                byte[] ba = mp.getContent();
                                if (mimeType.equals("image/png")) {
                                    content.append(buff.toString());
                                    Image img = Image.createImage(ba, 0, ba.length);
                                    ImageItem ii = new ImageItem(contentLocation,img,
                                        Item.LAYOUT_NEWLINE_AFTER, contentLocation);
                                    content.append(ii);
                                } else {
                                    buff.append(new String(ba));
                                    StringItem si = new StringItem("Part", buff.toString());
                                    si.setLayout(Item.LAYOUT_NEWLINE_AFTER);
                                    content.append(si);
                                }
                            }
                        }
                    }
                    display.setCurrent(content);
                }
            } catch (IOException e) {
                e.printStackTrace();
            }
        }
    public void pauseApp() {
```

```
                done = true;
                thread = null;
                resumeScreen = display.getCurrent();        }
    public void destroyApp(boolean unconditional) {
            done = true;
            thread = null;
            if (mmsconn != null) {
                try {
                    mmsconn.close();
                } catch (IOException e) {    }
            }
    }
    public void commandAction(Command c, Displayable s) {
            try {
                if (c == CMD_EXIT || c == Alert.DISMISS_COMMAND) {
                    destroyApp(false);
                    notifyDestroyed();
                }
            } catch (Exception ex) {    ex.printStackTrace(); }
        }
}
```

程序开始运行界面如图 10-26 所示。

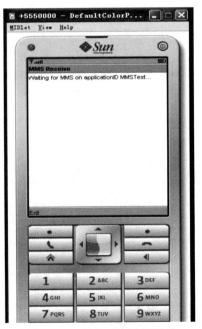

图 10-26　程序开始运行界面

打开多媒体发送 WMA 控制台，如图 10-27 所示，设置"Header"选项卡：主题为"Picture"，应用程序 ID 为"MMSTest"，发送到手机号为"5550000"。点击"Parts"选项卡，设置发送内容，如图 10-28 所示。

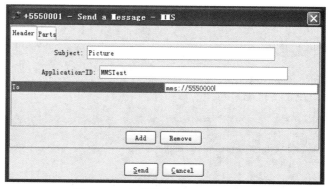

图 10-27　多媒体发送 WMA 控制台

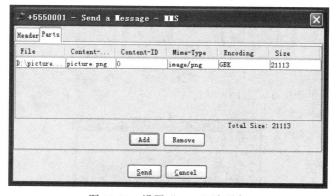

图 10-28　设置"Parts"选项卡

单击"Add"按钮，选择发送内容的图片。点击"Send"按钮，完成发送。这时在控制台上看到如图 10-29 所示内容，表示发送成功。

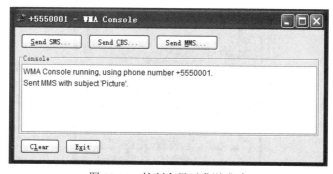

图 10-29　控制台显示发送成功

在模拟器手机中看到如图 10-30 所示界面，显示成功接收多媒体照片。

图 10-30　成功接收多媒体照片

10.5　本章小结

本章主要介绍无线消息传递的基本原理、WMA 的基本内容、使用无线消息 API 的基本方法、WTK 提供的 WMA 测试工具、SMS 消息的接收与发送，以及 CBS 消息的接收等内容。

短信是手机中不可或缺的功能，由于其具有灵活、价格低廉、无须接收端在线等优点，它已渗透到我们日常生活的各个方面。因此编写手机无线消息程序也是 Java ME 程序员必须掌握的技能之一。